MACMILLAN MASTER SERIES

Work Out

Mathematics

O Level and GCSE

The titles in this series

For examinations at 16+

Biology

Chemistry

Computer Studies

English Language

French

German

Mathematics

Physics

Principles of Accounts

Spanish

Statistics

For examinations at 'A' level

Applied Mathematics

Biology

Chemistry

English Literature

Physics

Pure Mathematics

Statistics

For examinations at college level

Mathematics for Economists

Operational Research

MACMILLAN
MASTER
SERIES

Work Out

Mathematics

'O' Level and GCSE

G.D. Buckwell

MACMILLAN

First published 1986
Reprinted 1986
Reprinted (with corrections) 1986

Published by
MACMILLAN EDUCATION LTD
Houndmills, Basingstoke, Hampshire RG21 2XS
and London
Companies and representatives
throughout the world

Typeset by TecSet Ltd,
Wallington, Surrey
Printed in Great Britain by The Bath Press, Avon

British Library Cataloguing in Publication Data
Buckwell, G. D.
Work out mathematics GCSE.——(Macmillan work
out series)——(Macmillan master series)
1. Mathematics——Examinations, questions, etc.
I. Title
510'.76   QA43
ISBN 0-333-39142-X

Contents

Acknowledgements   ix

    Groups responsible for Examinations at 16+   ix

Introduction   1

1   Set Theory

    1.1   Definition   2

    1.2   Equality   2

    1.3   Standard Sets   2

    1.4   Alternative Notation   2

    1.5   Universal Set, Complement   3

    1.6   Intersection, Null Set   3

    1.7   Union   3

    1.8   Subsets   4

    1.9   Venn Diagrams   4

    1.10 Number of Elements   4

    1.11 Sets of Points   5

    1.12 Logic   5

    1.13 De Morgan's Laws, Distributive Laws   6

2   The Number System

    2.1   The Number System   17

    2.2   Decimal Places and Significant Figures   18

    2.3   Number Bases   18

    2.4   Manipulation in Different Bases   20

    2.5   Fractions   21

    2.6   Fractions to Decimals   21

    2.7   Fraction as a Ratio   22

    2.8   Rate and Proportion   22

    2.9   Percentages   23

    2.10 Using Percentages, Profit and Loss   23

    2.11 Exchange Rates   24

    2.12 Map Scales   24

    2.13 Interest   25

    2.14 Large and Small Numbers, Standard Form   26

    2.15 Manipulation in Standard Form   28

    2.16 Approximations   28

    2.17 Surds   29

    2.18 Logarithms   30

    2.19 Rates   32

    2.20 Income Tax   33

    2.21 Wages   34

3   Algebra

3.1 Directed Numbers   39
3.2 Commutative, Associative, Distributive, Brackets   40
3.3 Substitution   41
3.4 Indices   42
3.5 Like and Unlike Terms   43
3.6 Removal of Brackets   43
3.7 Factors   44
3.8 LCM and HCF   46
3.9 LCM and HCF in Algebra   46
3.10 Algebraic Fractions   47
3.11 Forming Expressions   48
3.12 Linear Equations   49
3.13 Quadratic Equations   50
3.14 Simultaneous Equations (Linear)   51
3.15 Forming Equations   52
3.16 Changing the Subject of Formulae   54
3.17 Remainder Theorem   56
3.18 Variation and Proportion   57
3.19 Simultaneous Equations, One Linear,
     One Non-linear   58

4   The Metric System

4.1 Units   65
4.2 Changing Units in Area and Volume   65
4.3 Conversion of Units Involving Rates   66
4.4 Area   66
4.5 Volume   68
4.6 Density   68
4.7 Surface Area   68
4.8 Approximating Areas, Trapezium Rule   73
4.9 Radian Measure   74
4.10 Sectors and Segments   75
4.11 Speed, Velocity and Acceleration   77
4.12 Displacement–Time (or Distance–Time) Graphs   77
4.13 Velocity (Speed)–Time Graphs   77

5   Functions

5.1 Definition   85
5.2 Domain, Codomain, Range, Image   85
5.3 Composition of Functions   86
5.4 Inverse   86
5.5 Graphical Representation   87
5.6 The Equation of a Straight Line   89
5.7 Simple Coordinate Geometry (Distance, Mid-point)   90
5.8 Equation of a Circle   90
5.9 Graphical Solution of Equations   91
5.10 Curve Sketching   95
5.11 Variation and Proportion   96

6   Inequalities

6.1 Properties of $<$ and $\leqslant$ 102
6.2 Inequations   102
6.3 Errors   103
6.4 Regions   104

| | 6.5 | Linear Programming | 105 |
| | 6.6 | Locus | 109 |

7 Vectors

	7.1	Definition	113
	7.2	Multiplication by a Scalar	113
	7.3	Addition and Subtraction	114
	7.4	Length	114
	7.5	Equations	115
	7.6	Geometrical Proofs	115
	7.7	Locus Problems	118
	7.8	Velocity Triangles	118

8 Transformation Geometry

	8.1	Translation	124
	8.2	Reflection	124
	8.3	Rotation	125
	8.4	Glide Reflection	126
	8.5	Isometries	126
	8.6	Dilatation	126
	8.7	Shear	127
	8.8	Stretch	128
	8.9	Similarity and Congruence	128
	8.10	Geometrical Problems	128
	8.11	Area and Volume of Similar Figures	133

9 Matrices

	9.1	Array Storage	139
	9.2	Sum Difference, Multiplication by a Scalar	139
	9.3	Multiplication	140
	9.4	Algebraic Manipulation 1	140
	9.5	Identity	141
	9.6	Inverse Matrices	141
	9.7	Algebraic Manipulation 2	142
	9.8	Simultaneous Equations	142
	9.9	Networks and Route Matrices	144
	9.10	Incidence Matrices	145
	9.11	Euler's Theorem	146
	9.12	Traversability	146
	9.13	Transformations	147
	9.14	Combining Transformations	150

10 Straight Lines and Circles

	10.1	Parallel Lines	159
	10.2	The Triangle	159
	10.3	Polygons	160
	10.4	Regular Polygons	160
	10.5	Symmetry	161
	10.6	Geometric Properties of Simple Plane Figures	162
	10.7	The Circle (Definitions)	165
	10.8	The Properties of the Circle	165

11 Probability and Statistics

	11.1	Sample Space	173
	11.2	Definition of Probability	173
	11.3	Complement, Certain and Impossible Events	174
	11.4	Set Notation	174
	11.5	Combined Events, Tree Diagrams	176

		11.6	Statistical Diagrams	179
		11.7	Averages	181
		11.8	Frequency Tables	181
		11.9	Cumulative Frequency Curves (Ogives)	182
		11.10	Information from the Cumulative Frequency Curve	183
		11.11	Assumed Mean	183
		11.12	Histograms	184

12   Trigonometry

		12.1	Trigonometrical Ratios (Sine, Cosine, Tangent)	190
		12.2	Pythagoras' Theorem	192
		12.3	Relationship Between sin x, cos x and tan x	193
		12.4	Solution of Right-angled Triangles	194
		12.5	Special Triangles	195
		12.6	Angles of Elevation and Depression	197
		12.7	Bearings	197
		12.8	Reduction to Right-angled Triangles	199
		12.9	Perpendicular Height of a Triangle	201
		12.10	The Sine Rule	201
		12.11	Sine Rule (Ambiguous Case)	202
		12.12	The Cosine Rule	203
		12.13	Three-dimensional Problems	205
		12.14	Latitude and Longitude	207
		12.15	Velocity Triangles	209

13   Drawing and Construction

		13.1	Constructions Using Ruler and Compass	220
		13.2	Velocity Triangles	223
		13.3	Plans and Elevations	227
		13.4	Locus	229

14   Abstract Algebra

		14.1	Binary Operations	234
		14.2	Closure	234
		14.3	Identity	235
		14.4	Inverse	235
		14.5	Operation Tables	235
		14.6	Groups	236
		14.7	Symmetry Groups	240
		14.8	Modulo Arithmetic	241
		14.9	Subgroups	241
		14.10	Isomorphism	241

15   The Calculus

		15.1	Gradient of a Curve (Differentiation)	246
		15.2	Equation of the Tangent to a Curve	248
		15.3	Equation of a Normal to a Curve	248
		15.4	Rates of Change (Velocity and Acceleration)	249
		15.5	Integration	250
		15.6	Definite Integrals	251
		15.7	Area Under a Curve	251
		15.8	Volume of Revolution	253
		15.9	Maxima and Minima	254

| **Answers** | 262 |
| **Index** | 273 |

Acknowledgements

The author and publishers wish to thank the following who have kindly given permission for the use of copyright material:

The Associated Examining Board, the East Anglian Regional Examinations Board, the Joint Matriculation Board, the Northern Ireland Schools Examination Council, the Oxford and Cambridge Schools Examination Board, the Scottish Examination Board, the Southern Universities' Joint Board, the University of Cambridge Local Examinations Syndicate, the University of London School Examinations Board, the University of Oxford Delegacy of Local Examinations and the Welsh Joint Education Committee for questions from past examination papers.

Every effort has been made to trace all the copyright holders but if any have been inadvertently overlooked the publishers will be pleased to make the necessary arrangement at the first opportunity.

The University of London Entrance and Schools Examinations Council accepts no responsibility whatsoever for the accuracy or method in the answers given in this book to actual questions set by the London Board.

Acknowledgement is made to the Southern Universities' Joint Board for School Examinations for permission to use questions taken from their past papers but the Board is in no way responsible for answers that may be provided and they are solely the responsibility of the authors.

The Associated Examining Board, the University of Oxford Delegacy of Local Examinations, the University of Cambridge Local Examinations Syndicate, the Northern Ireland Schools Examination Council and the Scottish Examination Board wish to point out that worked examples included in the text are entirely the responsibility of the author and have neither been provided nor approved by the Board.

Groups responsible for Examinations at 16+

In the United Kingdom, examinations are administered by four examining groups and three examination boards. Syllabuses and examination papers can be ordered from the addresses given here.

Northern Examining Association

Joint Matriculation Board
　Publications available from:
John Sherratt and Son Ltd
78 Park Road
Altrincham, Cheshire WA14 5QQ [JMB]

Northern Regional Exam Board
Wheatfield Road, Westerhope
Newcastle upon Tyne NE5 5JZ

Yorkshire and Humberside Regional Exam Board
Scarsdale House
136 Derbyside Lane
Sheffield S8 8SE

North West Regional Exam Board
Orbit House, Albert Street
Eccles, Manchester M30 0WL

Associated Lancashire Schools Exam Board
12 Harter Street
Manchester M1 6HL

Midland Examining Group

University of Cambridge Local Examinations Syndicate
Syndicate Buildings, Hills Road
Cambridge CB1 2EU [UCLES]

Southern Universities' Joint Board
Cotham Road
Bristol BS6 6DD [SUJB]

West Midlands Regional Exam Board
Norfolk House, Smallbrook
Queensway, Birmingham B5 4NJ

Oxford and Cambridge Schools Examination Board
10 Trumpington Street
Cambridge CB2 1QB [O & C]

East Midlands Regional Exam Board
Robins Wood House, Robins Wood Road
Aspley, Nottingham NG8 3NR

London and East Anglian Group

University of London School Examinations Board
University of London Publications Office
52 Gordon Square
London WC1E 6EE [L]

East Anglian Regional Exam Board
The Lindens, Lexden Road
Colchester, Essex CO3 3RL

London Regional Exam Board
Lyon House
104 Wandsworth High Street
London SW18 4LF

Southern Examining Group

The Associated Examining Board
Stag Hill House
Guildford, Surrey GU2 5XJ [AEB]

Southern Regional Examining Board
Avondale House, 33 Carlton Crescent
Southampton, Hants SO9 4YL

University of Oxford Delegacy of Local Examinations
Ewert Place
Summertown,
Oxford OX2 7BZ [OLE]

South–Western Regional Examining Board
23–29 Marsh Street
Bristol BS1 4BP

Scottish Examination Board

Publications available from:
Robert Gibson & Sons (Glasgow) Ltd
17 Fitzroy Place, Glasgow G3 7SF [SEB]

Welsh Joint Education Committee
245 Western Avenue
Cardiff CF5 2YX [WJEC]

**Northern Ireland Schools Examination
    Council**
Examinations Office
Beechill House, Beechill Road
Belfast BT8 4RS [NISEC]

Introduction

Simply buying a book will not make you pass any examinations. However, by working from this book regularly and methodically you will improve your understanding of the key facts and have the opportunity to practice answering examination questions and so check your progress.

Each chapter is self-contained; so you should choose the topics you find particularly difficult, study the explanation and worked examples and then tackle the check test; the answers are at the back of the book. If you have problems with the check test, then go back through the earlier part of the chapter. Finally, to confirm what you have achieved, tackle the questions and exercises at the end of the chapter; again the answers are at the back.

Revision

You will probably revise more in mathematics than in any other subject. It is essential to know and understand the basic principles and be able to apply them accurately. Therefore it is important that you plan a realistic revision programme giving yourself time to practise working out problems in all the topics included in your examination. Start by obtaining the syllabus for the mathematics paper you are taking, from the appropriate examination board (addresses are given on pages ix and x).

Avoid panic by drawing up a revision programme which you can stick to. Just before the examination, go through the check tests included in each chapter of this book; this will improve your confidence and highlight any weak spots.

The Examination

This book provides a framework that will enable you to tackle an examination at the age of 16 – the GCSE. Pupils hoping for higher grades will gain most from the approach used here and the questions included are the type that will be met in papers 2 and 3 (where 3 papers are set) or papers 3 and 4 (where 4 papers are set).

Remember to keep a close watch on the time during the examination. If you simply cannot answer a question, move on to the next one without wasting time. At the end, if you have time, you can have another attempt at the questions you have missed.

1 Set Theory

1.1 Definition

A set is a clearly defined collection of distinct objects, which are called *members* or *elements* of the set. Sets will be denoted by capital letters $A, B, C, \ldots$ and the elements or members by lower-case letters $x, y, z, \ldots$. If x belongs to a set A, we write this $x \in A$. ($\in$ stands for 'belongs to', or 'is an element of' or 'is a member of'.)

If y does not belong to a set B, we write this $y \notin B$. Sometimes it is possible to list all the elements, for example $A = \{1, 3, 4, 7\}$, $B = \{a, b, x, y\}$. For more complex sets, see the notation used in section 1.4.

1.2 Equality

Two sets A and B can only be equal if A and B contain exactly the same elements. However, the order of listing elements is not important and so $\{1, 3, 4, 6\} = \{4, 1, 6, 3\}$.

1.3 Standard Sets

In mathematics, we are most frequently concerned with sets of numbers. In order to save repetition, certain letters are used for commonly used sets:

$\mathbb{N} = \{0, 1, 2, 3, 4, \ldots\}$
$\mathbb{Z} = \{0, \pm 1, \pm 2, \pm 3, \ldots\}$
$\mathbb{Z}^+ = \{1, 2, 3, 4, \ldots\}$
$\mathbb{Q}$ is the set of rational numbers (see Chapter 2).
$\mathbb{R}$ is the set of all real numbers from $-\infty$ to $+\infty$.

1.4 Alternative Notation

Listing each element of a set is often not possible, or convenient. We use the following notation:

$$A = \{x : x \in \mathbb{N} \text{ and } x \text{ is even}\}.$$

We read this as A equals the set of values of x such that x belongs to $\mathbb{N}$ and is even.

Hence, $A = \{0, 2, 4, 6, \ldots\}$. Clearly this notation is very suitable to represent an *infinite set*.

$$B = \{x : x \in \mathbb{R} \text{ and } 2x - 1 \leqslant 5\}.$$

Solving $2x - 1 \leqslant 5$ (see Chapter 6 if you are unsure about inequalities) gives $2x \leqslant 6, \therefore x \leqslant 3$.

This time, x can be any number less than or equal to 3. It could be represented by part of the number line from $-\infty$ to 3, but it is impossible to list the elements. The reader will find many more examples using this notation throughout the book.

1.5 Universal Set, Complement

The set of all possible elements that we wish to consider in any problem is called the *universal set*. It is denoted by $\mathcal{E}$. The universal set is particularly useful when considering the complement of a set A, denoted by A'. A' is the set of elements of $\mathcal{E}$ that do not belong to A.

$\therefore$ If $\qquad \mathcal{E} = \{1, 2, 3, 4, 5, 6, 7, 8\} \qquad$ and $\qquad A = \{x : x \text{ is odd}\}$,

then since $\qquad\qquad\qquad\qquad A = \{1, 3, 5, 7\}$

it follows that $\qquad\qquad\qquad\qquad A' = \{2, 4, 6, 8\}.$

1.6 Intersection, Null Set

The *intersection* of two sets A and B is the set of elements that belong to both set A and set B. The symbol $\cap$ is used.

Hence $\qquad\qquad\qquad A \cap B = \{x : x \in A \text{ and } x \in B\}$

(read as 'A intersection B').
For example, if

$$A = \{1, 3, 4, 5\}, \qquad B = \{2, 3, 5, 6\} \qquad \text{and} \qquad C = \{2, 6, 7\},$$

then:
(a) $A \cap B = \{3, 5\}$.
(b) $A \cap C$ has no elements. We write this

$$A \cap C = \{\ \} \qquad \text{or} \qquad A \cap C = \phi.$$

ϕ (pronounced 'phi') is known as the *null* set or the *empty* set. In this case $A \cap C = \phi$.

If $X \cap Y = \phi$ then the sets X and Y have no element in common. We say that X and Y are *disjoint* sets.

1.7 Union

The *union* of two sets A and B is the set of elements that belong to either set A or set B, or to both. The symbol $\cup$ is used.

Hence $\qquad\qquad A \cup B$ (read as 'A union B') $= \{x : x \in A \text{ or } x \in B\}$.

For example,

if $\qquad\qquad\qquad A = \{3, 4, 5, 6\} \qquad$ and $\qquad B = \{a, b, 2, 4, d\}$

then $\qquad\qquad\qquad\qquad A \cup B = \{2, 3, 4, 5, 6, a, b, d\}.$

3

1.8   Subsets

Sometimes we may want to indicate that all the elements of a set A belong to another set B. We say that A is a *subset* of B.

We write this $A \subset B$.

Note: Since all the elements of A belong to A, we can say that A is a subset of A. Hence $A \subset A$.

(Some books use the symbol $\subseteq$ to indicate that a set can be a subset of itself.)

1.9   Venn Diagrams

A problem can often be made much easier if it is represented by a Venn diagram. The universal set is represented by a rectangle and the sets are represented by closed curves, usually circles or ovals. Sets required can then be indicated by shading appropriate regions. The examples in Fig. 1.1 should indicate how this is done.

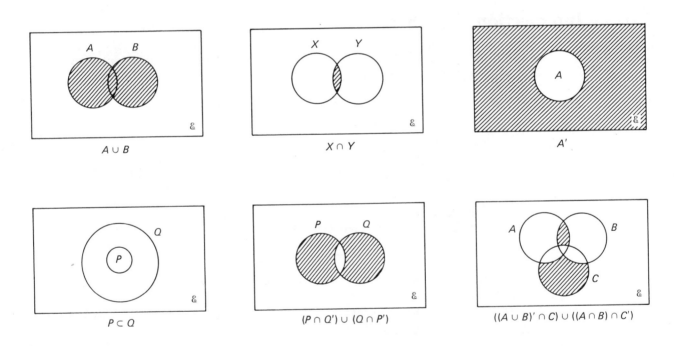

Fig. 1.1

1.10   Number of Elements

The number of elements in a set A is denoted by $n(A)$. For example,
(a) $A = \{3, 4, 7, 8, 10\}$, $n(A) = 5$,
(b) $B = \{x : 2 \leqslant x \leqslant 14$ and x is prime$\}$; since

$$B = \{2, 3, 5, 7, 11, 13\},$$

then $n(B) = 6$.

4

1.11 Sets of Points

Sets of points can often be conveniently represented using set notation.

$$A = \{(x, y) : y = 3x + 1\}.$$

This represents the set of points which satisfy $y = 3x + 1$, hence A is a straight line with equation $y = 3x + 1$.

Further examples of this can be found in Chapters 5 and 6. The following check exercise contains simple examples on the ideas covered so far.

Check 1

1 List the elements of the following sets:
  (a) $A = \{x : x \in \mathbb{N}, x \text{ is prime and } 8 < x < 30\}$.
  (b) $B = \{x : 4x + 3 = 9\}$.
  (c) $C = \{x : x \in \mathbb{N} \text{ and } 4x + 3 = 9\}$.
  (d) $D = \{(x \ y) : x + y = 8, x, y \in \mathbb{N}\}$.
  (e) $E = \{x : \sin x^\circ = \frac{1}{2}, 0 \leqslant x \leqslant 360\}$.

2 If $P = \{3, 4, 5, 6\}$, $Q = \{1, 4, 5, 7\}$, $R = \{1, 2, 4, 6\}$
   and the universal set $\& = \{1, 2, 3, 4, 5, 6, 7, 8\}$, find:
  (a) $P \cap Q$ (e) $(P' \cap Q) \cup R$
  (b) R' (f) $(P \cup Q') \cap R'$
  (c) $(P \cup Q) \cap R$ (g) $(P \cup Q) \cap (Q \cup R)$
  (d) $P \cup Q'$

3 Draw Venn diagrams to illustrate the following sets or relationships:
  (a) $(A \cup B) \cap C$ (d) $A' \cap (B \cup C)$ where $A \cap B = \phi$
  (b) $P \cap (Q' \cup R')$ (e) $(X \cap Y) \cup (Y \cap Z)$ where $X \cap Z = \phi$
  (c) $P \subset (Q \cup R)$

4 If $n(A) = 5$ and $n(B) = 8$, what are the greatest and least values of $n(A \cup B)$? Illustrate your answers with a Venn diagram.

5 If $n(X) = 8$, $n(Y) = 16$ and $n(X \cap Y) = 5$, what is $n(X \cup Y)$?

6 If $\& = \mathbb{Z}$, find:
  (a) $\{x : 3 \leqslant x \leqslant 8\} \cup \{x \ \ 5 < x < 9\}$,
  (b) $\{x : 2 < x < 7\} \cap \{x : 4 \leqslant x < 8\}$,
  (c) $\{x : x < 3\} \cup \{x : -8 \leqslant x \leqslant -5\}$.

1.12 Logic

Set notation is an extremely powerful tool when used to represent sentences and arguments in mathematical logic.

(a) If $P = \{\text{multiples of 4}\}$ and $Q = \{\text{multiples of 2}\}$ then the relationship $P \subset Q$ could be used to represent the statement, 'All multiples of 4 are divisible by 2.'

(b) If $A = \{\text{domestic animals}\}$ and $B = \{\text{whales}\}$ then the relationship $A \cap B = \phi$ could be used to represent the statement, 'A whale is not a domestic animal.'

(c) Consider the truth of the following argument:

> All racing cars are fast.
> Some fast cars expensive.
> Hence, there are some racing cars that are expensive.

There can be no doubt in reality, about the truth of the final statement, but look at the mathematics behind the argument.

Let $F = \{\text{fast cars}\}$, $E = \{\text{expensive cars}\}$ and $R = \{\text{racing cars}\}$.

The information given can be expressed in set notation by the two results $R \subset F$ and $F \cap E \neq \phi$.

If we now try to represent these results in a Venn diagram, we find that there are two possible diagrams satisfying these conditions. These can be seen in Fig. 1.2.

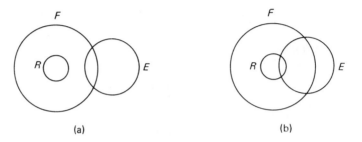

(a)   (b)

Fig. 1.2

Clearly, Fig. 1.2(a) does not satisfy the final conclusion. Hence, the final conclusion is false.

1.13   De Morgan's Laws, Distributive Laws

There are certain results which are true for any set A. These can be used to simplify results (see Worked Examples 1.1 and 1.5). They are:

(a) $A \cup A = A$;

(b) $A \cup A' = \mathcal{E}$;

(c) $A \cup \mathcal{E} = \mathcal{E}$;

(d) $A \cup \phi = A$;

(e) $A \cap A = A$;

(f) $A \cap A' = \phi$;

(g) $A \cap \mathcal{E} = A$;

(h) $A \cap \phi = \phi$.

De Morgan's laws:

(i) $(A \cup B)' = A' \cap B'$;

(j) $(A \cap B)' = A' \cup B'$.

The proofs of (i) and (j) given below illustrate another way of using Venn diagrams.

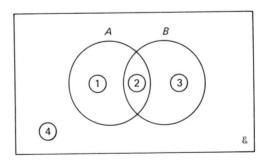

Fig. 1.3

Each region in Fig. 1.3 has been numbered 1, 2, 3 or 4.

$$(A \cup B)' = (① \cup ② \cup ③)' = ④$$
$$A' \cap B' = (③ \cup ④) \cap (① \cup ④) = ④$$
$$\therefore (A \cup B)' = A' \cap B'.$$

Proof of the other result is similar.

The distributive laws:
(k) $A \cup (B \cap C) = (A \cup B) \cap (A \cup C)$;
(l) $A \cap (B \cup C) = (A \cap B) \cup (A \cap C)$.

Worked Example 1.1

For the sets A and B, we define $A * B$ as the set of elements which are in A or B but not in both. The set $A * B$ is shaded in the Venn diagram, Fig. 1.4.

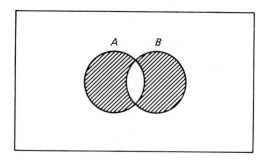

Fig. 1.4

Write down, without using *, sets which are equivalent to
(a) $A * A$ (c) $(A * B) \cup A$
(b) $A * A'$ (d) $(A * B) \cup B$ [London]

Solution

(a) We can write $A * B = (A \cap B') \cup (A' \cap B)$,
hence $A * A = (A \cap A') \cup (A' \cap A) = \phi \cup \phi = \phi$.
(b) $A * A = (A \cap (A')') \cup (A' \cap A')$
$= (A \cap A) \cup A' = A \cup A' = \mathscr{E}$.
(c) $(A * B) \cup A$ can be seen most clearly if a Venn diagram is used. If all of A is shaded in addition to $A * B$, then the whole of $A \cup B$ is shaded,
$\therefore (A * B) \cup A = A \cup B$.
(d) As in part (c), it can be seen that
$(A * B) \cup B = A \cup B$.

Worked Example 1.2

In a local election, 3025 constituents voted for a number of candidates, 800 voted for candidate A only, 850 voted for candidate B only, and 630 voted for candidate C only. 425 voted for A and B, 120 voted for A and C and 57 voted for B and C.

If 537 voted for none of the candidates A, B and C, how many voted for all three candidates?

Solution

This question is best solved by a Venn diagram.

If the number of people who voted for all three candidates is denoted by x, then the number in each region will be as in Fig. 1.5.

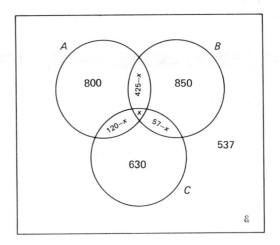

Fig. 1.5

Since the total number of people represented in the diagram is 3025, then

$$800 + (425 - x) + 850 + (120 - x) + x + (57 - x)$$
$$+ 630 + 537 = 3025.$$
$$\therefore\ 3419 - 2x\ = 3025.$$

Hence $x = 197$.

$\therefore$ 197 people voted for all three candidates.

Worked Example 1.3

Two sets P and Q are such that $P \cup Q = P$. A third set R is such that $R \cap Q = \phi$. Which of the following statements *must* be true?

(a) $P \cap R \neq \phi$;   (b) $P \cap Q \cap R = \phi$;   (c) $Q \subset P \cup R$.

Solution

The main difficulty with this type of problem is that the given information can be represented by more than one Venn diagram. All possibilities must be considered before attempting an answer.

$P \cup Q = P$ means that Q is a subset of P.

$R \cap Q = \phi$ only means that R does not overlap Q. The three possibilities are shown in Fig. 1.6.

(a) Fig. 1.6(c) shows that $P \cap R \neq \phi$ can be false.

(b) In each diagram $P \cap Q \cap R = \phi$, hence the statement is true.

(c) Since $P \cup R$ contains all of P, and Q is a subset of P, then $Q \subset P \cup R$ must be true.

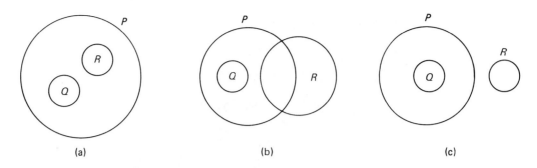

(a)   (b)   (c)

Fig. 1.6

Worked Example 1.4

If $\mathscr{E}$ = {quadrilaterals}, A = {rectangles}, B = {squares},
  C = {rhombuses} and D = {parallelograms},
draw a Venn diagram to show the relationship between the sets.

Solution

A square is not a rectangle, hence $A \cap B = \phi$.
A square is a special type of rhombus, hence $B \subset C$.
A rectangle cannot be a rhombus, hence $A \cap C = \phi$.
A, B and C are all subsets of D.
The relationship can then be shown as in Fig. 1.7.

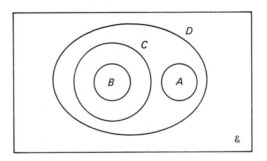

Fig. 1.7

Worked Example 1.5

In this question, P and Q are sets, and $P * Q$ means $P \cap Q'$. In a Venn diagram, $P * Q$ could be shown by the shaded region of Fig. 1.8.

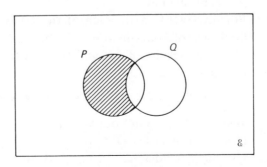

Fig. 1.8

Draw three separate Venn diagrams and shade the areas which represent

(a) $P' * Q$ (b) $P * Q'$ (c) $(P * Q) \cap P$ [L]

Solution (see Fig. 1.9(a)–(c).)

(a) $P' * Q = P' \cap Q' = (P \cup Q)'$.
(b) $P * Q' = P \cap (Q')' = P \cap Q$.
(c) $(P * Q) \cap P = (P \cap Q') \cap P$
$$= P \cap (Q' \cap P) = P \cap (P \cap Q')$$
$$= (P \cap P) \cap Q' = P \cap Q'.$$

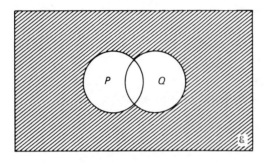

Fig. 1.9(a)

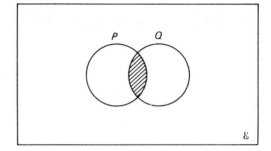

Fig. 1.9(b)

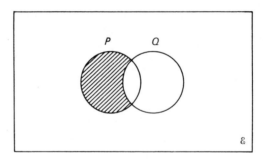

Fig. 1.9(c)

Worked Example 1.6

Binary operations $\circ$ and $*$ are defined such that, for any two sets A and B, $A \circ B = A \cap B', A * B = (A \circ B) \cup (B \circ A)$.

(a) Using three separate Venn diagrams, show each of the sets $A \cap B, A \circ B$ and $A * B$ by shading.

(b) For any set A, write each of the sets $\phi * A$ and $A * A$ as a single set. What does each of your answers indicate?

(c) Using the results of (b) and the fact that $*$ is associative, show that for sets P, R and X

$$P * X = P * R \Rightarrow X = R.$$

Justify each step in your working.
(*Hint*: Consider $P * (P * X)$.)

(d) Use the Venn diagram in Fig. 1.10 to explain why

$$J \cup K = J \cup L \nRightarrow K = L.$$

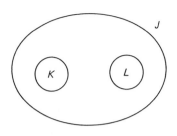

Fig. 1.10

(e) Explain, with the aid of another Venn diagram, why

$$C \cap D = C \cap F \quad \Rightarrow \quad D = F.$$ [AEB]

Solution

(a) See Fig. 1.11.

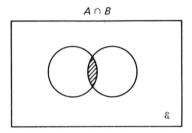 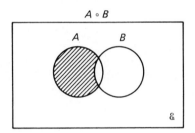 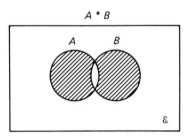

Fig. 1.11

(b) $\phi * A = (\phi \circ A) \cup (A \circ \phi)$,

  but $\phi \circ A = \phi \cap A' = \phi$ and $A \circ \phi = A \cap \phi' = A \cap \& = A$.

  $\therefore \phi * A = \phi \cup A = A$.

  This indicates that ϕ is the identity element under $*$.

  $A * A = (A \circ A) \cup (A \circ A) = (A \cap A') \cup (A \cap A') = \phi$.

  This indicates that A is self-inverse under $*$.

  (*Note*: See Chapter 14 for further details.)

(c) $P * X = P * R$,

  $\therefore P * (P * X) = P * (P * R)$.

  Since $*$ is associative, it follows that

  $(P * P) * X = (P * P) * R$.

  Using the results in (b),

  $\phi * X = \phi * R$,

  $\therefore X = R$.

(d) Since K is a subset of J, $J \cup K = J$, and also since L is a subset of J, $J \cup L = J$. Therefore $J \cup K = J \cup L$ but clearly $K \neq L$.

(e) The diagram in Fig. 1.12 shows that in each case, $C \cap D = C \cap F = C$ but clearly $D \neq F$.

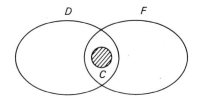

Fig. 1.12

Exercise 1

1 If $A = \{x : 3 \leqslant 2x < 18\}$, then $n(A)$ equals:
  A 7 **B** 8 **C** 9 **D** none of these
2 If $A \cup (B \cap C) = B$, then it follows that:
  A $A \cap C = \phi$ **B** $B = C$ **C** $A \subset B$ **D** $A \subset C \subset B$
3 If $A \cap B' = \{3, 4, 5\}$, $B \cap A' = \{1, 6, 8\}$ and $\& = \{1, 2, 3, 4, 5, 6, 7, 8, 9, 10\}$, then $n(A \cap B)$ equals:
  A 3 **B** 4 **C** 6 **D** none of these
4 If $X \subset Y$ and $Y \cap Z = \phi$ then $Y \cup (X \cap Z)$ equals:
  A X **B** Y **C** Z **D** none of these
5 If $P = \{x : 1 < x + 4 \leqslant 8\}$ and $Q = \{x : x < 4\}$ then $P \cap Q$ equals:
  A $\{x : 1 < x < 4\}$ **B** $\{x : x \geqslant -3\}$ **C** $\{x : -3 < x < 4\}$
  D $\{x : -3 \leqslant x < 4\}$.
6 If $n(A) = 9$, $n(B) = 8$ and $n(A \cap B') = 5$, then $n(A \cap B)$ equals:
  A 4 **B** 3 **C** 5 **D** none of these
7 In the Venn diagram in Fig. 1.13, the shaded area represents:
  A $P' \cup Q$ **B** $P' \cap Q$ **C** $P' \cup Q'$ **D** $P' \cap Q'$ [AEB 1980]

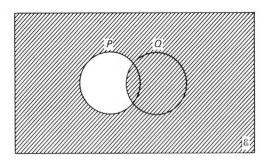

Fig. 1.13

8 A set S contains 8 elements and a set T contains 13 elements. The least possible number of elements in $S \cap T$ is:
  A 0 **B** 5 **C** 8 **D** 13   [AEB 1981]
9 Two sets P and Q are such that $P \cup Q = \&$ and $P \cap Q \neq \phi$. It follows that $P \cup Q'$ simplifies to:
  A Q' **B** Q **C** P **D** P'

10 If R = {rich people} and T = {teachers}, then the statement, 'Some teachers are not rich people', can be represented by:

A $T \subset R$ **B** $T \subset R'$ **C** $T \cap R' = \phi$ **D** $T \cap R' \neq \phi$.

11 In the universal set {3, 5, 7, 9, 11, 13, 15}, the sets X and Y are given by $X = \{5, 9, 11\}$ and $Y = \{5, 9, 13, 15\}$. Find:

(a) $X' \cap Y'$ (b) $(X \cup Y')'$

12 The universal set $\&$ is defined

    $\& = $ {even numbers from 2 to 60 inclusive},

and P, Q and R are subsets of $\&$ such that

    $P = $ {multiples of 3},

    $Q = $ {perfect squares},

    $R = $ {numbers, the sum of whose digits is 8}.

Find: (a) $P \cap Q$ (b) $P \cap Q'$ (c) $(R \cup Q) \cap P'$

13. In a survey of 149 households about an evening's television viewing, 25 watched BBC1 and ITV only, 53 watched BBC2 and BBC1 only, 26 watched BBC2 only and 27 watched ITV only. If 8 had not watched the television, and nobody watched all three, how many had watched ITV? Assume nobody had watched BBC1 only.

14 Newspapers are delivered to 40 houses. Each house receives either one copy of *The Times* or one copy of *The Gazette* or one copy of each. In all, 26 copies of *The Times* and 24 copies of *The Gazette* are delivered. Find the number of houses which receive *The Times* only.

15 Use set notation to describe the sets shaded in Fig. 1.14.

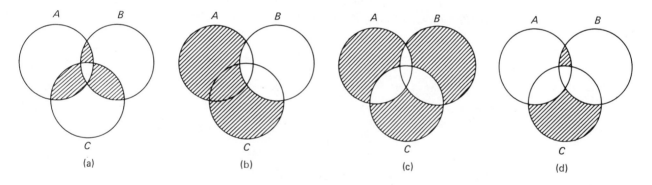

(a)   (b)   (c)   (d)

Fig. 1.14

16 The universal set is the set of all plane quadrilaterals.

    The following subsets are defined:

$A = $ {quadrilaterals with exactly one line of symmetry};

$B = $ {quadrilaterals with four sides equal};

$C = $ {quadrilaterals with exactly two internal angles of 90°};

$D = $ {squares};

$\phi = $ the empty set.

(a) Which one of the above sets is equal to $A \cap B$?

(b) Draw a quadrilateral, q, for which $q \in A \cap C$. What is the name for this type of quadrilateral?

(c) Explain why a quadrilateral which belongs to set B cannot also belong to set C.

(d) Draw a Venn diagram to represent sets A, B and C, illustrating your answers to (a), (b) and (c).

(e) Illustrate set D in an appropriate position on your diagram.

(f) Draw a quadrilateral which belongs to $A' \cap C$.   [OLE]

17 (a) Draw a single Venn diagram to illustrate the relations between the following sets:

P = {parallelograms},

Q = {quadrilaterals},

R = {rectangles},

S = {squares},

Z = {quadrilaterals having *one and only one* pair of parallel sides}.

State which one of the sets P, R, S, Z, ϕ is equal to

(i) $R \cap S$, (ii) $P \cap Z$.

(b) The 89 members of the fifth form all belong to one or more of the Chess Club, the Debating Society and the Jazz Club.

Denoting these sets by C, D and J respectively, it is known that 20 pupils belong to C only, 15 to J only and 12 to D only. Given that $n(C \cap J) = 18$, $n(C \cap D) = 20$ and $n(D \cap J) = 16$, calculate (a) $n(C \cap J \cap D)$, (b) $n(D')$.

[L]

18 If $\& $ = {vehicles}, F = {fast cars}, and R = {red cars}, express in good English

(a) $R \cap F = \phi$, (b) $R \cap F \neq \phi$, (c) $R \cap F' \neq \phi$.

19 The universal set $\&$ consists of the positive integers less than 30. P and Q are subsets, P containing those members of $\&$ which are multiples of 5 and Q those members of $\&$ which are multiples of 3. List the members of the sets:

$$P, Q, P \cap Q', P' \cap Q, P \cup Q, P' \cup Q' \text{ and } (P \cap Q') \cup (P' \cap Q).$$

Represent the sets $\&$, P and Q by a Venn diagram and shade in the region or regions representing $(P \cap Q') \cup (P' \cap Q)$.

20 The universal set is the set of positive integers which are less than or equal to 40. The sets M, N and P are defined as follows:

$M = \{x : x \text{ is a multiple of } 3\}$,

$N = \{x : x \text{ is a factor of } 36\}$,

$P = \{x : x \text{ is a prime number}\}$.

Find: (a) M, N, P (b) $M \cap N \cap P$ (c) $(M \cup N) \cap P'$

21 $\&$ is the set of the first twenty natural numbers,

i.e. $\& = \{1, 2, \ldots, 20\}$

List the members of the following subsets:

(a) $A = \{t : 10 < 4t - 5 \leqslant 43\}$,

(b) $B = \{y : y \text{ is a prime number}, y > 1\}$.

(c) $C = \{z : z \text{ is a factor of } 20, z > 1\}$.

Draw a Venn diagram showing the relationship between $\&$, A, B and C, writing each of the members of $\&$ in the appropriate region.

List the members of the following sets:

(d) $A' \cap B$ (e) $(A \cup B)'$ (f) $B' \cap C'$

22 Of the 125 members of a youth club, 46 like table tennis (T), 38 like darts (D) and 63 like snooker (S). 13 like table tennis and darts, 12 like darts and snooker and 16 like table tennis and snooker. All members like at least one of the three sports.

Illustrate this information in a Venn diagram and find:

(a) the number of members who like all three.

(b) the number of members who like only darts.

23 Some of the results of a survey amongst 70 fifth-year pupils are shown in the Venn diagram, Fig. 1.15.

$\&$ = {pupils questioned in the survey},

C = {pupils who went to the cinema},

T = {pupils who went to the theatre},

P = {pupils who went to a pop concert}.

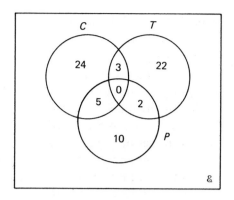

Fig. 1.15

Calculate the number of pupils
(a) who had been to the cinema,
(b) in $C \cup P$,
(c) in $C \cap T$, explaining the meaning of your answer,
(d) in $P \cup (C \cap T)$,
(e) in $(C \cup T \cup P)'$, explaining the meaning of your answer.

[AEB]

24 The universal set is the set of positive integers less than or equal to N.
$A = \{x : x$ is even$\}$,
$B = \{y : y$ is exactly divisible by 7$\}$,
$C = \{z : z$ is a multiple of 5$\}$.
(a) Find $A \cap B \cap C$.
(b) If $n(A \cap B) = 5$, what is the largest possible value of N?

25 In a school where 80 students study sciences in the sixth form, 40 study chemistry, 46 study physics and 31 study biology. 20 study physics and biology, and the number who study physics and chemistry is three times the number who study chemistry and biology. If only 3 students take all three sciences, how many study only physics?

26 The Venn diagram, Fig. 1.16, concerns the numbers of players in a ladies' sports club.
$B = \{$ladies who play badminton$\}$,
$H = \{$ladies who play hockey$\}$,
$T = \{$ladies who play tennis$\}$.

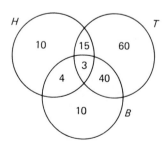

Fig. 1.16

The numbers shown indicate the number of ladies in each set. Find the number of ladies who play (a) badminton only; (b) all three games; (c) hockey and tennis but not badminton; (d) two of the three games but not all three.

[AEB]

27 Given that X and Y are two subsets of $\mathscr{E}$ and that $n(X \cap Y) \neq 0$, draw Venn diagrams to illustrate:

(a) $(X \cup Y) \cap (X \cap Y')'$

(b) $(X' \cap Y') \cup (X' \cap Y)$

28 (a) $\mathscr{E}$ is the set of natural numbers $\{1, 2, 3, 4, 5, \ldots\}$.

List the members of the following sub-sets:

(i) $\{x: 3x - 1 = 8\}$ (ii) $\{y : 2y + 1 < 8\}$ (iii) $\{t : t - 1 \leqslant 4\}$

(b) $\mathscr{E}$ is the set of the first 50 natural numbers $\{1, 2, 3, \ldots, 50\}$.

A is the set $\{x : x \in \mathscr{E}$ and x is an even number$\}$,

B is the set $\{y : y \in \mathscr{E}$ and y is a square of a natural number$\}$,

C is the set $\{x : z \in \mathscr{E}$ and $z < 20\}$.

Write down the members of the sets

(i) $A \cap B$ (ii) $B \cap C'$ (iii) $A \cup B$ (iv) $(A \cap B) \cap (B \cap C')$

29 (a) In a survey of shoppers it was found that 60 per cent were pensioners and 30 per cent were male. If 60 per cent of the pensioners were female, what percentage of the shoppers were female non-pensioners? Show the information on a Venn diagram.

(b) A survey of cars in a car-park revealed that the most popular optional extras were radios, clocks and sunroofs.

6 of the cars had radios and sunroofs, 14 had radios and clocks and 5 had clocks and sunroofs but did not have radios.

20 had sunroofs, 60 had radios and 22 had clocks but neither radios nor sunroofs.

(i) If x cars had clocks, radios and sunroofs, complete a Venn diagram showing all the above information.

(ii) If there were 120 cars in the car-park, write down, on the Venn diagram, how many cars had none of these optional extras (answer in terms of x).

(iii) If 19 cars had exactly 2 of these optional extras, write down an equation and solve it to find x.   [SUJB]

2 The Number System

2.1 The Number System

The numbers we use in mathematics can be classified into various sets, some of which are subsets of others. The definitions are as follows:

{Positive integers} = {1, 2, 3, 4, 5, . . .} = {natural numbers}.
{Negative integers} = {. . . −5, −4, −3, −2, −1}.
{Integers} = {. . . −5, −4, −3, −2, −1, 0, 1, 2, 3, 4, . . .}.
{Rational numbers} = {$x : x$ can be expressed as a fraction}.
{Irrational numbers} = {$x : x$ cannot be expressed as a fraction}.
{Prime numbers} = {$x : x$ has only two different factors, 1 and itself}.
= {2, 3, 5, 7, 11, 13, . . .}.
{Square numbers} = {1, 4, 9, 16, 25, . . .}.
There are many others.

Notes: (1) A terminating decimal is rational, e.g.
$$3.12 = \frac{312}{100} = \frac{78}{25}.$$
(2) A non-terminating repeating decimal is rational, e.g.
$3.\dot{1}\dot{2} = 3.121212 \ldots = 3\frac{4}{33}.$
(3) π is irrational.
(4) An integer is rational, since it can be expressed as a fraction, e.g.
$$6 = \frac{6}{1}.$$
(5) Numbers such as $\sqrt{2}, \sqrt{3}$, which are not whole numbers, are irrational numbers. If left like this without working them out, they are referred to as *surds*. The manipulation of these is shown in section 2.17.

Worked Example 2.1

State whether the following values of x are rational or irrational:

(a) $x = 12.4$;   (b) $x^2 = 8$;   (c) $\pi x + 4\pi = 0$;   (d) $x^2 - 5x + 6 = 0$;
(e) $2.\dot{3}$.

Solution

(a) Since $12.4 = \frac{124}{10} = \frac{62}{5}$ then x is *rational*.
(b) If $x^2 = 8$, $x = \pm\sqrt{8}$. But this is not an exact number, hence x is *irrational*.
(c) Although π appears in this question, we can factorize the left-hand side.
     Hence $\pi (x + 4) = 0, \therefore x = -4$ and hence x is *rational*.
(d) The equation $x^2 - 5x + 6 = 0$ can be factorized as $(x - 3)(x - 2) = 0$. Hence $x = 2$ or $x = 3$, therefore x is *rational*.
(e) $0.\dot{3}$ as a fraction is $\frac{1}{3}$, hence $x = 2.\dot{3} = 2\frac{1}{3}$ and x is *rational*.

2.2 Decimal Places and Significant Figures

The importance of stating the correct number of digits in any calculation cannot be emphasized too much with the advent of the electronic calculator. Too many digits in any answer are usually irrelevant. All answers in this book are given correct to three significant figures unless stated otherwise. Also, unless stated otherwise, the digit 5 will be rounded up.

Examples using decimal places:

The following numbers are corrected to two decimal places (abbreviation d.p.):

$$84.739 = 84.74 \text{ (2 d.p.)}$$
$$6.104 = 6.10 \text{ (2 d.p.)} \leftarrow \textbf{The zero must be included.}$$
$$0.0199 = 0.02 \text{ (2 d.p.)}$$
$$0.999 = 1.00 \text{ (2 d.p.)}$$

The following numbers are shown corrected to three significant figures (abbreviation s.f.):

$$128.4 = 128 \quad \text{(3 s.f.)}$$
$$6.093 = 6.09 \quad \text{(3 s.f.)}$$
$$0.01498 = 0.0150 \text{ (3 s.f.)}$$

Significant figures are not counted until the first non-zero digit is reached reading from the left. In this case the 1 is the first digit and the 9 is the third.

Check 2

1 Write the following numbers correct to 2 d.p.:
   (a) 8.689   (b) 26.134   (c) 0.094   (d) 0.099   (e) 2.019
   (f) 2.190   (g) 9.999

2 Write the following numbers correct to 3 s.f.:
   (a) 86.41   (b) 186.9   (c) 0.09   (d) 1.999   (e) 169.86
   (f) 3.004   (g) 8694   (h) 9999

2.3 Number Bases

Written in full, the denary number 4186 would be four thousand, one hundred and eighty-six,

or
$$4186 = 4 \times 1000 + 1 \times 100 + 8 \times 10 + 6$$
$$= 4 \times 10^3 + 1 \times 10^2 + 8 \times 10^1 + 6 \times 10^0.$$

The number 10 in this case is referred to as the *base* of the number.

The number 69 can be written in two ways:

$$69 = 6 \times 10^1 + 9 \times 10^0,$$

or
$$69 = 1 \times 2^6 + 0 \times 2^5 + 0 \times 2^4 + 0 \times 2^3 + 1 \times 2^2 + 0 \times 2^1 + 1 \times 2^0.$$

In the first case base 10 is used (denary); in the second, base 2 is used (binary).

In binary, $69 = 1000101_2 \longleftarrow$ denotes the base.

Clearly, any number can be used for the base.

Check 3

The following numbers should be converted into denary to refresh your mind:

(a) 136_8 (b) 214_8 (c) 2150_8 (d) 1011_2 (e) 111101_2
(f) 100011_2 (g) 3140_5 (h) 2415_6 (i) 6135_7 (j) 33333_5.

The process of changing the base of a number is illustrated by the following example:

Worked Example 2.2

(a) Change 683 into binary.
(b) Change 63_8 into base 3.
(c) Change 1011101_2 into base 8 (often referred to as octal).

Solution

The method usually involves a continuous division process writing down the remainders.

(a)

```
                              ┌──── Remainders
                      2 ) 683   ▼
   New base ──────▶   2 ) 341  r 1
                      2 ) 170  r 1
                      2 )  85  r 0
                      2 )  42  r 1
                      2 )  21  r 0
                      2 )  10  r 1
                      2 )   5  r 0
                      2 )   2  r 1
                      2 )   1  r 0
  Stop when ─────────▶      0  r 1
  this number                  ▲
  is zero                      └──── Read upwards for the answer
```

$\therefore$ $683 = 1010101011_2$.

(b) If changing from one base to another, usually you have to go via base 10.

$63_8 = 6 \times 8^1 + 3 \times 8^0 = 48 + 3 = 51$.

Change 51 to base 3:

```
        3 ) 51
        3 ) 17  r 0
        3 )  5  r 2
        3 )  1  r 2
             0  r 1
```

$\therefore$ $63_8 = 1220_3$.

(c) Base 8 to base 2 can be changed in a quicker fashion.

$$1 \mid 011 \mid 101$$
$$\uparrow \qquad \uparrow \qquad \uparrow$$
$$1 \qquad 3 \qquad 5$$

Divide the binary number *into* groups of three digits reading from the *right*. Change each of these 3-digit (or less at the front) numbers into a denary number.

Hence $1011101_2 = 135_8$.

To change an octal number into a binary number, simply change each digit to binary reading from the right and write down the binary numbers in order.

Quickly try the following:

Check 4

1 Change the following numbers into binary:
   (a) 48   (b) 86   (c) 17   (d) 1496   (e) 212_3 (f) 543_6
   (g) 76_8 (h) 134_8.
2 Change the following binary numbers into octal numbers:
   (a) 1011   (b) 10111   (c) 11011011   (d) 1010101.

2.4 Manipulation in Different Bases

Some knowledge of addition, multiplication etc. is assumed. The following example should clarify any problems.

Worked Example 2.3

Evaluate in the given bases:
(a) $10111_2 + 111_2$ (b) $4341_8 - 556_8$ (c) $210_3 \times 122_3$
(d) $10010000_2 \div 1100_2$.

Solution

(a)
```
  10111
+   111
 ------
  11110
   1 1 1
```

(b)
```
   1011
   3239
   4341
  - 556
  -----
   3563
```
Note: if 8 is carried this changes the 1 to a 9.

(c)
```
           210
         × 122
        ------
         21000
Here    /11200
2 × 2 = 11/  1120
base 3   ------
         111020
            1 1
```

(d)
```
               1100
            312
1100  10010000
      1100
      ----
       1100
       1100
       ----
        000
```

You should now try the following:

Check 5

Evaluate the following questions in the stated bases:

1 $1101_2 + 111_2$	2 $1011_2 \times 11_2$	3 $212_3 + 122_3$
4 $650_8 + 573_8$	5 $11011_2 - 111_2$	6 $(47_8)^2 \times 12_8$
7 $24415_6 \div 35_6$	8 $651_8 \div 21_8$	9 $1331_8 \div 33_8$
10 $1000_2 - 1111_2$	11 $1111_2 + 1111_2$	12 $412_6 - 354_6$
13 $2201_3 - 222_3$	14 $27_8 \times 63_8$	15 $35_6 \times 21_6$

2.5 Fractions

The following example should refresh your memory about the rules of working with fractions.

Worked Example 2.4

Evaluate the following:
(a) $\frac{1}{4} + \frac{2}{3}$ (b) $4\frac{3}{5} - 2\frac{5}{6}$ (c) $\frac{4}{9} \times 1\frac{1}{4}$ (d) $3\frac{1}{2} \div 2\frac{1}{4}$

Solution

(a) $\frac{1}{4} + \frac{2}{3} = \frac{3}{12} + \frac{8}{12} = \frac{11}{12}$.

(b) $4\frac{3}{5} - 2\frac{5}{6} = \dfrac{23}{5} - \dfrac{17}{6} = \dfrac{138}{30} - \dfrac{85}{30} = \dfrac{53}{30}$
$$= 1\frac{23}{30}.$$

(c) $\frac{4}{9} \times 1\frac{1}{4} = \dfrac{\overset{1}{\cancel{4}}}{9} \times \dfrac{5}{\underset{1}{\cancel{4}}} = \frac{5}{9}$.

(d) $3\frac{1}{2} \div 2\frac{1}{4} = \dfrac{7}{2} \div \dfrac{9}{4} = \dfrac{7}{\underset{2}{\cancel{2}}} \times \dfrac{\overset{2}{\cancel{4}}}{9} = \dfrac{14}{9} = 1\frac{5}{9}$.

'Turn the second upside down'

Carry out the following check test *without* using a calculator.

Check 6

1 $\frac{3}{5} + \frac{3}{4}$	2 $\frac{1}{3} + \frac{1}{5}$	3 $\frac{1}{2} + \frac{1}{3} + \frac{1}{4}$	4 $\frac{2}{5} - \frac{1}{4}$
5 $\frac{3}{5} - \frac{1}{2}$	6 $\frac{3}{8} - \frac{1}{6}$	7 $\frac{2}{5} \times \frac{5}{8}$	8 $\frac{3}{4} \times 1\frac{1}{2}$
9 $\frac{3}{5} \div \frac{2}{5}$	10 $1\frac{1}{2} \div 2\frac{1}{2}$	11 $2\frac{1}{3} + 4\frac{1}{2}$	12 $(\frac{3}{8} + \frac{1}{4}) \times 2\frac{1}{2}$

13 $(\frac{2}{5} \times \frac{1}{4}) \div \frac{1}{2}$ 14 $\dfrac{\frac{2}{5} + \frac{1}{2}}{\frac{1}{10} + \frac{1}{4}}$ 15 $(\frac{1}{3} - \frac{1}{4}) \times (2\frac{3}{4} - 1\frac{1}{5})$

16 $(4\frac{1}{2} \div 2\frac{1}{4}) \times 1\frac{1}{2}$ 17 $(\frac{3}{4} + 2\frac{1}{5}) \div \frac{1}{10}$ 18 $(\frac{3}{5})^2 \div \frac{1}{4}$

19 $3\frac{1}{3} \div (\frac{1}{4} + \frac{1}{8})$ 20 $\frac{1}{5} \div (\frac{3}{4} \times \frac{4}{5})$

2.6 Fractions to Decimals

It was mentioned in section 2.1 that a fraction (irrational number) could sometimes be expressed as an infinite repeating fraction. The following examples show

how to change a fraction to a decimal, using the process of short division.

$$\text{(a) } \tfrac{4}{25}: \qquad 25\overline{)4.00}^{\,0.16}$$

$$= 0.16$$

$$\text{(b) } \tfrac{5}{14}: \qquad 14\overline{)5.00000000}^{\,0.35714285\ldots}$$

$$= 0.3\dot{5}7142\dot{8}.$$

Now try the following.

Check 7

Convert the following fractions into decimals. Avoid using a calculator.

1 $\tfrac{3}{5}$ 2 $\tfrac{3}{25}$ 3 $\tfrac{5}{8}$ 4 $\tfrac{5}{7}$ 5 $\tfrac{2}{11}$ 6 $\tfrac{13}{25}$ 7 $\tfrac{5}{18}$ 8 $\tfrac{7}{19}$
9 $\tfrac{7}{16}$ 10 $\tfrac{17}{48}$

2.7 Fraction as a Ratio

If Jane has £5 and David has £8, we could express the amount of money that Jane has as a fraction of the amount David has; it would be $\tfrac{5}{8}$. However, more commonly we express the amounts as a ratio:

$$\text{Jane : David} = £5 : £8 \text{ or just } 5 : 8.$$

Ratios behave much like fractions and can be simplified.

$$\text{(a) } 40 : 280 = 4 : 28 = 1 : 7;$$
$$\text{(b) } 2\tfrac{1}{2} : 4\tfrac{1}{4} = 5 : 8\tfrac{1}{2} = 10 : 17.$$

Quantities can only be compared using ratios if they are measured in the same units.

2.8 Rate and Proportion

Material is priced at £8 per metre. This idea of some quantity per unit quantity is called a *rate*. Many rate problems are about speed and acceleration, where the unit is time. It follows that 2 m costs £16 and 3 m costs £24. The quantity of price is proportional to the length in metres. If you double the length you double the price. If you treble the length you treble the price.

We could express the cost £C of L m as

$$C = 8L.$$

In general, if y is *directly proportional* to x then y is a multiple of x,

$$\therefore y = kx;$$

k is called the *constant of proportionality*.

More examples of proportion will be given in later chapters. (See particularly section 3.18.)

2.9 Percentages

A percentage is a fraction where the denominator equals 100. For example,

$$\tfrac{3}{4} = \tfrac{75}{100} = 75\% \, .$$

A less obvious example is

$$\frac{5}{12\frac{1}{2}} = \frac{5 \times 8}{12\frac{1}{2} \times 8} = \frac{40}{100} = 40\%.$$

To change a fraction to a percentage, simply multiply the fraction by 100.
To change a percentage to a fraction, divide the percentage by 100.

Check 8

1 Change into percentages:
   (a) $\tfrac{2}{5}$ (b) 2.79   (c) $\tfrac{1}{6}$ (d) 0.0375   (e) $\tfrac{1}{200}$ (f) $\tfrac{11}{20}$
   (g) $\tfrac{3}{2}$ (h) $\tfrac{49}{500}$
2 Change into a fraction in its simplest form:
   (a) $33\tfrac{1}{3}\%$ (b) 72%   (c) $12\tfrac{1}{2}\%$ (d) 0.6%   (e) 12%
   (f) 36.5%

2.10 Using Percentages, Profit and Loss

To find a percentage of a quantity, proceed as in the following examples.

Find 8% of £25

Money problems are often simplified by the fact that 1 per cent of £1 is 1p.
   In this case, 1 per cent of £25 is 25p.

$$\therefore 8\% \text{ is } 8 \times 25p = £2,$$

or 8% is $\tfrac{8}{100} \times £25 = £2$.

Express 7 cm as a Percentage of 3 m

Percentages can only be used to change ratios if the quantities have the same units.

$$7 \text{ cm} : 3 \text{ m} = 7 : 300$$

can be written as a percentage:

$$\frac{7}{300} \times 100 = 2\tfrac{1}{3}\%.$$

In problems on profit and loss, changes or errors, the formula used is

$$\text{Percentage profit (loss, change, error)} = \frac{\text{Profit (loss, change, error)}}{\text{Cost (original value)}} \times 100.$$

The danger in using this formula is shown in the following example.

Worked Example 2.5

A dealer buys a stamp collection and sells it for £2700, making a 35% profit. Find the cost of the collection.

Solution

Note that it is wrong to find 35 per cent of £2700 and subtract this amount from £2700.

In fact, £2700 = 135% of the cost,

$$\therefore \frac{£2700}{135} \times 100 = 100\%,$$

∴ cost is £2000.

2.11   Exchange Rates

The ideas used in money changing are those of ratio or proportion.

For example, if £1 = $1.22, find
(a) how many dollars you would get for £8.40,
(b) how many pounds you would get for $8.40.

(a) Since $£1 = \$1.22,$

multiplying each side by 8.4:

$$£1 \times 8.4 = \$1.22 \times 8.4$$
$$\therefore £8.40 = \$10.248.$$

(b) In the reverse process, reduce the number of dollars to one:

$$£1 = \$1.22.$$

Divide each side by 1.22:

$$£\frac{1}{1.22} = \$1,$$
$$\therefore £0.8197 = \$1.$$

Multiply each side by 8.4:

$$£0.8197 \times 8.4 = \$8.4 \times 1,$$
$$\therefore £6.885 = \$8.40.$$

2.12   Map Scales

(a) Lengths

The scale of a map is expressed as a ratio, for example 1 : 50 000. This means that any distance on the map represents 50 000 times that distance in reality. Hence 1 cm = 50 000 cm. Changed to km, 50 000 cm is 0.5 km. Hence, on the map, 1 cm = 0.5 km.

(b) Areas

In order to find the area represented by a map, the map ratio must be squared. (See section 8.11 on similar figures.)

In the above example, the ratio of areas is therefore $1^2 : 50\,000^2$, i.e. $1 : 2.5 \times 10^9$. Hence on the map 1 cm^2 represents 2.5×10^9 cm^2. To change to km^2, divide by $100\,000^2 = 10^{10}$.

$$\therefore 1 \text{ cm}^2 = 0.25 \text{ km}^2.$$

2.13 Interest

(a) Simple Interest

If a sum of money called the *principal* is invested at a given interest rate per annum, it earns *interest* depending on the time invested.

If P = principal invested,

I = interest earned,

R = rate (percentage) per annum,

T = time in years,

then $I = \dfrac{PRT}{100}$.

In practice, most interest is calculated as *compound* interest but *simple* interest is often used if the period invested is less than one year.

Worked Example 2.6

If 250 is invested at 12 per cent p.a. simple interest for 1 year 3 months, find the total amount of money at the end of this time.

Solution

Using the formula $I = \dfrac{PRT}{100}$,

$$P = 250, \qquad R = 12, \qquad T = 1\tfrac{1}{4},$$

$$I = \frac{250 \times 12 \times 1.25}{100} = 37.5.$$

$$\therefore \quad \text{Total} = \pounds 250 + 37.5 = \pounds 287.50.$$

(b) Compound Interest

If the interest is added at the end of each year, hence increasing the principal, the interest is compound.

Worked Example 2.7

If £800 is invested at $7\frac{1}{2}$ per cent compound interest for a period of 3 years, find the value of the investment at the end of that time.

Solution

The interest for each year is added to the principal at the beginning of each year, giving the new principal for the next year. We can proceed as follows:

First year: principal = £800,

$$\text{interest} = £800 \times \frac{7.5}{100} = £60.$$

Second year: principal = £800 + 60 = £860,

$$\text{interest} = £860 \times \frac{7.5}{100} = £64.50.$$

Third year: principal = £860 + 64.50 = £924.50,

$$\text{interest} = £924.50 \times \frac{7.5}{100} = £69.34.$$

∴ The value of the investment = £924.50 + 69.34 = £993.84.
The formula

$$\text{Total} = P\left(1 + \frac{R}{100}\right)^n$$

can be used to find the capital at the end of n years.
In the above example,

$$\text{Capital} = £800 \left(1 + \frac{7.5}{100}\right)^3 = £993.84.$$

2.14 Large and Small Numbers, Standard Form

(a) Large Numbers

You are no doubt aware that, using indices, we can write for example:

$$10\,000 = 10^4.$$

Using this notation enables large numbers to be written in a more convenient way, using standard form. For example:

(a) 4.8×10^3 means $4.8 \times 1000 = 4800$,
(b) 28.5×10^4 means $28.5 \times 10\,000 = 285\,000$.

In these two examples (a) *is in standard form*, (b) *is not*.

> To write a number in standard form it is written as $A \times 10^n$ and we must have that
>
> $$1 \leqslant A < 10.$$

To change a number into standard form, proceed as in the following example.

To Change 4 860 000 into Standard Form

Place a decimal point after the first digit (from the left), i.e. 4.860 000.

Count the number of digits after the decimal point including the zeros; in this case there are six: 4.860 000.

$$\underbrace{\qquad\qquad}_{6 \text{ digits}}$$

The answer is 4.86×10^6.

(b) Small Numbers

Now consider the situation for very small numbers.

We can write

$$\tfrac{1}{10} = 10^{-1}, \tfrac{1}{100} = 10^{-2} \text{ etc.}$$

(See section 3.4 for more work on indices.)

Therefore, $0.086 = 8.6 \div 100 = 8.6 \times \tfrac{1}{100}$.

Hence $0.086 = 8.6 \times 10^{-2}$.

$$\underbrace{\qquad}_{2 \text{ zeros}}$$

You should notice that the negative power of 10 is the number of zeros, including one in front of the decimal point, before the first non-zero digit.

(c) Using a Calculator

To enter a number in standard form on most calculators, proceed as follows.

To enter 3.95×10^4:

Press:	*Display:*
[3.95] [EXP] [4]	[3.95 04]

To enter 6.95×10^{-3}:

Press:	*Display:*
[6.95] [EXP] [±] [3]	[6.95 − 03]

When a calculation gives an answer containing too many digits for the calculator, it automatically puts the answer into standard form.

Carry out the following simple questions without a calculator.

Check 9

1 Change into standard form:
   (a) 486   (b) 6930   (c) 857 000
   (d) 0.01   (e) 0.096   (f) 0.008 31   (g) 0.0909
2 Change from standard form:
   (a) 4.5×10^4 (b) 3.84×10^3 (c) 8.01×10^6 (d) 8.5×10^{-2}
   (e) 6.95×10^{-4} (f) 8.801×10^{-6}

2.15 Manipulation in Standard Form

(a) Addition or Subtraction

Numbers are better changed from standard form first.

$$2.86 \times 10^4 + 4.3 \times 10^2 = 28\,600 + 430 = 29\,030 = 2.903 \times 10^4.$$

(b) Multiplication

$$\begin{aligned}
(3.9 \times 10^4) \times (4.6 \times 10^5) &= 3.9 \times 4.6 \times 10^4 \times 10^5 \\
&= 3.9 \times 4.6 \times 10^9 . \leftarrow(4 + 5; \text{see section 3.4.}) \\
&= 17.94 \times 10^9 .
\end{aligned}$$

Note that this is not standard form; in standard form the answer would be 1.794×10^{10}.

(c) Division

$$(4.52 \times 10^5) \div (8.86 \times 10^{-2})$$

$$= \frac{4.52 \times 10^5}{8.86 \times 10^{-2}} = \frac{4.52}{8.86} \times \frac{10^5}{10^{-2}}$$

$$= 0.510 \, (3\,\text{s.f.}) \times 10^7 . \leftarrow(5 - -2 = 7; \text{see section 3.4.})$$

Once again, this is not in standard form; in standard form the answer would be 5.1×10^6.

Most calculators will carry out calculations directly on numbers entered in standard form, but answers will often be given in decimal form.

2.16 Approximations

The ability to approximate is a useful skill in that it enables you to check an answer quickly without working out the exact answer. Basic methods involve some form of guessing, or use of standard form. Consider the following example.

Find correct to one significant figure, the value of

(a) $\dfrac{84 \times 959}{63 \times 0.96}$ (b) $\dfrac{6.93 \times (0.87)^2}{4.23}$ (c) $\dfrac{869 \times 4965}{0.091 \times 9513}$

As a general rule, if you make a number on the top line smaller (or larger), make a number on the bottom line smaller (or larger).

(a) $\dfrac{84 \times 959}{63 \times 0.96} \approx \dfrac{80 \times 960}{60 \times 0.96}$ ($\approx$ means approximately equal)

$$\approx \frac{8\emptyset \times 1000}{6\emptyset \times 1} = \frac{8000}{6} \approx 1000.$$

The correct answer is 1332, therefore our approximation is correct to 1 s.f.

(b) $\dfrac{6.93 \times (0.87)^2}{4.23} \approx \dfrac{7 \times 1^2}{5} \approx 1.$

The correct answer is 1.24.

(c) $\dfrac{869 \times 4965}{0.091 \times 9513} \approx \dfrac{9 \times 10^2 \times 5 \times 10^3}{9 \times 10^{-2} \times 10^4}$

$$= \dfrac{45 \times 10^5}{9 \times 10^2} = 5 \times 10^3 = 5000.$$

The correct answer is 4984.

Try approximating the following calculations, correct to 1 s.f., checking your answer with a calculator.

Check 10

1 $\dfrac{86.3 \times (0.8)^2}{60}$

2 $\dfrac{4.3 \times 0.08}{6.5 \times 7.1}$

3 $\dfrac{(39)^2 \times 43^2}{81 \times 964}$

4 $7.3 \div (8.9 \times 2.7^2)$

5 $\dfrac{6.97 + 8.53}{4.21 + 7.35}$

6 $\dfrac{4395 \times (0.07)^2}{0.19}$

7 $\dfrac{63 \div (8.6 \times 10^{-5})}{6.47 \div 3.85}$

8 $\dfrac{(0.09)^2 \div (6.3 \times 10^{-5})}{7.4 \times 2951}$

2.17 Surds

The fact that $\sqrt{3}$ is an irrational number was mentioned in section 2.1. It is often convenient to manipulate quantities like this called *surds*, without working them out.

(a) $\sqrt{3} \times \sqrt{3} = 3.$

(b) $\dfrac{1}{\sqrt{3}} = \dfrac{1}{\sqrt{3}} \times \dfrac{\sqrt{3}}{\sqrt{3}} = \dfrac{\sqrt{3}}{3}.$

(c) $\sqrt{12} = \sqrt{4 \times 3} = 2\sqrt{3}$ since $\sqrt{4} = 2.$

(d) $\sqrt{75} + \sqrt{12} = \sqrt{25 \times 3} + 2\sqrt{3} = 5\sqrt{3} + 2\sqrt{3} = 7\sqrt{3}$

(e) $\dfrac{1}{\sqrt{3} + \sqrt{2}} = \dfrac{1 \times (\sqrt{3} - \sqrt{2})}{(\sqrt{3} + \sqrt{2})(\sqrt{3} - \sqrt{2})}$

(this is called rationalizing the denominator)

$$= \dfrac{\sqrt{3} - \sqrt{2}}{(\sqrt{3})^2 - (\sqrt{2})^2} \quad \text{(see section 3.7d)}$$

$$= \dfrac{\sqrt{3} - \sqrt{2}}{3 - 2} = \sqrt{3} - \sqrt{2}.$$

Check 11

1 Simplify the following:

    (a) $\sqrt{18} + 3\sqrt{2}$ (b) $\sqrt{24} + \sqrt{150}$ (c) $\sqrt{18} - \sqrt{8}$

    (d) $\sqrt{90} + \sqrt{40}$ (e) $2\sqrt{32} - \sqrt{128}$ (f) $\sqrt{63} - \sqrt{28}$

    (g) $\sqrt{99} - \sqrt{176}$ (h) $\sqrt{28} + \sqrt{7} + \sqrt{63}$

    (i) $\sqrt{12} - \sqrt{48} + \sqrt{98} + \sqrt{18}$ (j) $\sqrt{45} + \sqrt{20}$

2 Express the following with a rational denominator:

    (a) $\dfrac{1}{\sqrt{2}}$ (b) $\dfrac{1}{\sqrt{125}}$ (c) $\dfrac{1}{\sqrt{8} - \sqrt{2}}$ (d) $\dfrac{2}{\sqrt{8}}$ (e) $\dfrac{5}{\sqrt{10}}$

    (f) $\dfrac{3}{\sqrt{45}}$ (g) $\dfrac{1}{\sqrt{7} - \sqrt{2}}$

2.18 Logarithms

The use of logarithm tables for calculation has now become obsolete with the advent of the electronic calculator. However, the logarithm function is still of considerable importance in advanced work.

> Consider the expression $y = 10^x$.
> If $x = 2$, $y = 10^2 = 100$.
> We say that 2 is the logarithm of 100 using base ten. Any number could be used for the base.

Figure 2.1 shows the shape of the graph of $y = 10^x$. It cannot be drawn accurately, because the values of y become very large.

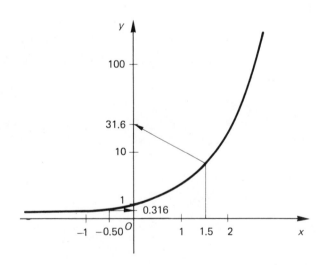

Fig. 2.1

If it had been plotted accurately, we could read off $x = 1.5$, $y = 31.6$.
This is abbreviated to $1.5 = \log_{10} 31.6$, i.e. $31.6 = 10^{1.5}$.
Also, if $x = -0.5$, $y = 0.316$, hence $0.316 = 10^{-0.5}$.

In fact, we could find the logarithm of any number and plot a graph of $y = \log_{10} x$, which would look like the graph in Fig. 2.2. It is in fact the *reflection* of the curve $y = 10^x$ in the line $y = x$.

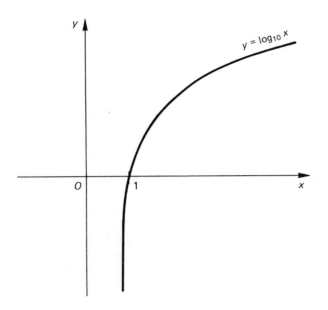

Fig. 2.2

$\log_{10} x$ is usually written $\log x$.

On the calculator, there is also a button labelled ln. This is a logarithm using base 2.718 (denoted by e). The function $\ln x$ ($\log_e x$) is necessary when trying to find $\int \frac{1}{x}\, dx$. See section 15.5.

In fact, $\int \frac{1}{x}\, dx = \log_e x + C$.

There are three simple rules for manipulating logarithms, in any base:

(1) $$\log M + \log N = \log MN$$

(2) $$\log M - \log N = \log \frac{M}{N}$$

(3) $$n \log M = \log M^n$$

These are similar to the rules for indices, and it is important to state that a logarithm is really a *power*.

Worked Example 2.8

Use logarithms to solve the equation $4^x = 7$.

Solution

Clearly x is not a whole number. When two numbers are equal, their logarithms are equal,

$$\therefore \log 4^x = \log 7.$$

Using rule (3), $\qquad x \log 4 = \log 7$

$$\therefore x = \frac{\log 7}{\log 4} = \frac{0.8451}{0.6020} \quad \text{using the calculator}$$

$$= 1.404 \text{ (3 d.p.).}$$

Worked Example 2.9

Solve the equation

$$\log (x + 1) + \log (x + 2) = 3.$$

Solution

Using rule (1), the left-hand side simplifies to give

$$\log (x + 1) (x + 2) = 3.$$

Assuming the logarithms are to base 10,

$$3 = \log 1000,$$
$$\therefore \log (x + 1) (x + 2) = \log 1000,$$
$$\therefore (x + 1) (x + 2) = 1000,$$
i.e. $\qquad\qquad x^2 + 3x + 2 = 1000,$
$$x^2 + 3x - 998 = 0.$$

The quadratic formula must be used here; see section 3.13 (b).

$$\therefore x = \frac{-3 \pm \sqrt{9 + 3992}}{2} = 30.13 \qquad \text{or} \qquad -33.13.$$

From the graph of $\log x$, it can be seen that x cannot be negative, and since $\log (x + 1)$ and $\log (x + 2)$ appear in the original equation, it follows that x cannot equal -33.13.

Hence, solution is $x = 30.13$.

2.19 Rates

Every property is given a *rateable value* by the local authority. The *rate* is levied at so much in the £1 of rateable value. Rates are also levied separately for water.

Worked Example 2.10

The rateable value of a house used to be £260, and the annual amount paid in rates was £218.40. The new rateable value is £285 and the new rate is 91p in the pound. The old water rate was 8 per cent of the old rateable value and is now 9 per cent of the new rateable value.

(a) Find the yearly increase in rates payable, and the percentage increase in the rate per £1.

(b) Find the increased amount due for the water rate.

Solution

(a) The new rate payable is 285 × 91p
$$= £259.35.$$
The increase in rates = £259.35 − £218.40 = £40.95.

Since the old rate of £218.40 was obtained by a rate of x pence per £1 on £260, it follows that 260 × x = 218.4,

$$\therefore \quad x = \frac{218.4}{260} = 0.84.$$

The percentage increase in the rate = $\frac{(91 - 84)}{84} \times 100 = 8\frac{1}{3}\%.$

(b) The old water rate = $\frac{8}{100} \times £260 = £20.80.$

The new water rate = $\frac{9}{100} \times £285 = £25.65.$

The increase in water rate = £25.65 − £20.80
$$= £4.85.$$

If the total rateable value of an area is say £400 000, then an increase of 1p in the pound for rates would bring in £400 000 × $\frac{1}{100}$ = £4000.

This is called a *penny rate*.

2.20 Income Tax

Income Tax is paid by most people on their income after certain deductions have been made. These deductions determine a *tax code number*, the use of which is illustrated by the following example.

Worked Example 2.11

A man has a tax code number of 245. If he earns £8240 a year, find the amount of tax he pays each week when the tax rate is 30 per cent.

Solution

To find the allowances from the tax code, multiply the code by 10.

Total amount earned = £8240.
Total allowances = 245 × 10 = £2450.
∴ Taxable income = £8240 − £2450 = £5790.

∴ Tax paid = $\frac{30}{100} \times £5790 = £1737.$

∴ Weekly tax = $\frac{£1737}{52} = £33.40.$

For higher earners, the rate becomes staggered.

2.21 Wages

These are best illustrated by the following simple example.

Worked Example 2.12

An apprentice earns £1.80 an hour. His basic week is 35 hours. If he works overtime, he is paid 'time-and-a-half' except for weekends, when he is paid double time. Calculate the amount he earns in a week if he works 39 hours from Monday to Friday, Saturday 09.30 hours to 14.00 hours and $2\frac{1}{4}$ hours on Sunday.

Solution

In a basic week, the apprentice earns 35 × £1.80 = £63. Since he works 39 hours before Friday, we have 4 hours overtime at 'time-and-a-half', i.e. at £1.80 + 90p per hour:

$$4 \times £2.70 = £10.80.$$

On Saturday he works $4\frac{1}{2}$ hours, on Sunday $2\frac{1}{4}$ hours. The total is $6\frac{3}{4}$ at 'double time', i.e. at £3.60 an hour:

$$6\frac{3}{4} \times £3.60 = £24.30.$$

His total wage for the week is

$$£63 + 10.8 + 24.3 = £98.10.$$

Worked Example 2.13

A dealer buys a piece of antique furniture and later sells it for a 75 per cent profit. If he sells it for £1505, find how much he paid for it.

Solution

It is important not to take 75 per cent from the selling price. This is what happens:

$$75\% \text{ of } £1505 = \frac{75}{100} \times £1505 = £1128.75.$$

If we now assume he bought it for £1128.75, let us find 75 per cent profit on this amount:

$$\frac{75}{100} \times £1128.75 = £846.56.$$

Selling price = £1128.75 + 846.56 = £1975.31, a long way out.
   The correct method is as follows:
The selling price is 175 per cent of the cost price.

$$\therefore £1505 = 175\%;$$

$$\div 175: \frac{£1505}{175} = 1\%;$$

$$\times 100: \frac{£1505}{£175} \times 100 = 100\%.$$

Hence purchase price = £860.

Exercise 2

1 If x is a rational number and y is an integer, then:

    A xy cannot be an integer   **B** xy can be irrational

    C $x + y$ is rational   **D** $x + y$ is an integer

2 If $a : b = 2 : 5$ and $b : c = 7 : 9$, then $a : c$ equals:

    A $18 : 35$ **B** $7 : 22\frac{1}{2}$ **C** $9 : 14$ **D**   an indeterminate value

3 If $(2^x)^2 = 4$, then x equals:

    A   1   **B**   2   **C**   3   **D**   4

4 The scale of a map is $1 : 25\,000$. The ratio of the area of a field to its area on the map is:

    A $2.5 \times 10^4 : 1$ **B** $6.25 \times 10^{16} : 1$ **C** $1 : 6.25 \times 10^8$

    D $6.25 \times 10^8 : 1$

5 If y is directly proportional to the square of x, and $y = 4$ when $x = 2$, it follows that when $y = 16$, x equals:

    A   4   **B**   256   **C**   8   **D** $\frac{1}{2}$

6 The value of $\dfrac{18.1 \times 0.705}{630}$ estimated correct to one significant figure is:

    A   0.02   **B**   0.2   **C**   0.002   **D**   0.003

7 A dealer sold a television set for £88, and made a 10 per cent profit. It follows that he bought it for:

    A   £79.20   **B**   £80   **C**   £78   **D**   none of these

8 If $x = 2 \times 10^4$ and $y = 3 \times 10^3$, then $2x + 3y$ equals:

    A 1.3×10^8 **B** 9.4×10^4 **C** 1.3×10^4 **D** 4.9×10^4

9 If $1227_n \div 21_n = 47_n$, then n equals:

    A   7   **B**   8   **C**   9   **D**   11

10 If n is an integer, then $n^2 + n + 1$ is:

    A   always odd   **B**   always even   **C**   a prime number

    D   sometimes divisible by 5

11 Which of the following has the largest numerical value?

    A $\sqrt{10}$ **B** $\sqrt{100}$ **C** $\dfrac{1}{\sqrt{0.01}}$ **D** $\dfrac{1}{0.01}$ [AEB]

12 The average of $\frac{1}{4}$ and $\frac{1}{3}$ is:

    A $\frac{1}{6}$ **B** $\frac{2}{7}$ **C** $\frac{7}{24}$ **D** $\frac{7}{12}$

13 The scale of a map is $1 : 50\,000$. The distance, in km, represented by 40 cm on the map is:

    A   2   **B**   0.5   **C**   20   **D**   5

14 A tenant paid a monthly rent of £121 twelve times a year and a half-yearly rate of £58.40. Find to the nearest penny how much he paid for rent and rates per week (52 week year).

15 A town council have decided to increase the local rate by 6p in the pound. The total rateable value is £4 800 000. What will be the increase in income from the rates?

    A householder finds out that his yearly rates will increase from £246 to £264. What is the rateable value of the house, and what was the original rate before the 6p increase?

16 (a) The following subtraction is not in base ten. Find the base, and determine the missing number x.

$$\begin{array}{r} 1324 \\ \underline{x42} \\ 572 \end{array}$$

    (b) The number 1234_n is divided by n^2. Write down the remainder in base n.

                                                                    [OLE]

17 Express the numbers 6286, 0.0104 and 2947 in standard form and use your answers showing clearly your working, to find an approximation to $\frac{6286 \times 0.0104}{2947}$ correct to one significant figure.

18 In 26 hours, a car mechanic earns £67.08. Find out how much he will earn, at the same rate of pay per hour, in 34 hours.

19 Calculate the following, giving your answers in each case as a binary number:
  (a) $2^5 + 2^3 + 2$ (b) $1001_2 - 11_2$
  (c) $657_8 + 24_8$ (d) $111_2 \times 101_2$

20 Show that for all positive integers n,
  (a) 169_n is a perfect square,
  (b) 156_n cannot be a prime number.

21 What base has been used in the following calculations?
  (a) $41 \times 35 = 2355$ (b) $33 + 123 = 222$
  (c) $1605 \div 65 = 21$

22 If $x = a + 3b$ and $y = 4a - 3b$, find a and b, given that when $x = 1.2 \times 10^5$, $y = 3.6 \times 10^6$.

23 It is given that y is directly proportional to the square of x. If $y = 2$ when $x = 3$, find the relationship that connects x and y. Find the possible values of y if $y = 8x$.

24 A line AB is divided into two parts, AC and CB, such that $AC : CB = 3 : 5$. A further point D divides CB into two parts such that $CD : DB = n : 1$. If $AD : DB = 19 : 5$, find n.

25 Given that $\sqrt{2} = 1.41$, $\sqrt{3} = 1.73$, $\sqrt{10} = 3.16$, evaluate as accurately as possible:
  (a) $\sqrt{6}$ (b) $\sqrt{0.02}$ (c) $\sqrt{500}$

26 In 1981, the owner of a coach business found that his running costs were £456 860 and his receipts from fares were £430 640. For 1982 he dispensed with uneconomic routes, which he hoped would reduce his costs by 14.5 per cent, and he estimated that the fall in receipts from fares would be 6.5 per cent. What trading profit did he expect in 1982? In fact, fuel price increases meant that the actual reduction in running costs was only 12 per cent and he made a trading loss of £10 154. Calculate the actual percentage fall in receipts from fares. (Give your answers correct to four significant figures.)   [OLE]

27 The total rateable valuation of the property in a borough council area is £7 500 000.
  (a) How much is raised by levying a penny rate?

  To provide for an increase in expenditure next year, the finance committee of the council estimates that an extra £240 000 will have to be demanded from the borough ratepayers.
  (b) (i) Calculate by how much the rate in the £ will have to be increased next year.
  (ii) Determine the increase in rates which will have to be paid by Mr Smith, whose property has a rateable valuation of £540.

  This year's rate is 116.8p in the £. Next year Mr Jones will receive a rate demand for £276.
  (c) (i) What is the rateable valuation of his property?
  (ii) What will be the percentage increase in the rates demanded of Mr Jones?
      [NISEC]

28 A company which hires out caravans provides holidays in France. The table gives the total cost of caravan hire plus boat fares for two adults and their caravan:

	June	July/August
four-berth caravan		
1 week	£110	£122
2 weeks	£143	£160
3 weeks	£175	£199
six-berth caravan		
1 week	£121	£135
2 weeks	£161	£181
3 weeks	£203	£231

If more than two adults travel with the caravan, there is a charge of £16 for each extra adult and £8 for each child. This charge does not depend on the length of the holiday.

(a) (i) Mr and Mrs Morris hire a four-berth caravan for two weeks in June. Calculate the cost if they take Mrs Morris's mother with them on their holiday.

(ii) Mr and Mrs Jones hire a four-berth caravan for three weeks in August and take one child and his grandmother with them. Calculate the total cost.

(iii) Mr and Mrs Hill hire a six-berth caravan for two weeks in July and take their three children with them. Calculate the total cost.

(b) In addition to the cost in (a) (iii) the Hill family expect to spend £90 on petrol, £110 on entertainment, £140 on food, £40 on other items, and also to pay a camping charge of £3 for each of 13 nights. Calculate the total expected cost (including the fares and caravan hire) of their holiday.

(c) If the costs in (a) (ii) were increased by 10 per cent, calculate the new total cost for caravan hire and fares for the Jones family.   [UCLES]

29 The cost, £6664, of making up a private road is shared between the council, the house-owners in the road and the house-owners in an adjoining road. The council pays one half of the total costs. The eight house-owners in the adjoining road each pay equal shares for access rights totalling one-eighth of the total costs. Each house-owner in the private road pays in proportion to the length of the frontage of the house owned. The seven houses in the road have frontages of 14, 15, 10.5, 12.5, 21.5, 24.5 and 49 metres. Calculate *exactly*:

(a) the amount paid by each owner in the adjoining road for access rights;

(b) the amount paid by the owner occupying the house with the shortest frontage;

(c) the percentage of the total costs paid by the owner of the house with the 24.5 m frontage.

The house owners in the private road agree to share equally the costs of extra work, tree planting etc. amounting to £385.

(d) Calculate the total amount now paid by the owner of the house with the 10.5 m frontage.   [L]

30 A $\frac{1}{2}$p coin is 16 mm in diameter. These coins are stamped out of a strip of metal 16 mm wide and 300 cm long. What is the maximum number of coins that can be stamped from this strip? The waste metal is then melted and made into a strip 16 mm wide again, and of the same thickness. How many $\frac{1}{2}$p coins can be cut from this strip? (Area of a circle = πr^2, $\pi = 3.14$.)

31 The speed of light is 3×10^8 m/s. Find the distance that light travels in one year (365 days). Give your answer in the form $A \times 10^n$, where A is correct to

three significant figures. It is found that two stars are 25 light-years away. What is the distance between these two stars? Express your answer in the form $A \times 10^n$.

32 An American woman spent a holiday in Paris, Rome and London. In each place, the total cost of her visit consisted of the payment of her hotel bill together with all other extra expenses.

In Paris: the cost at her hotel was 164 francs per day; she stayed for 9 days; the ratio of her extra expenses to her hotel bill was 3 : 5.

In Rome: her hotel bill was four times her extra expenses; the total cost of her visit was 425 000 lire.

In London: her total cost amounted to 100 dollars more than her total cost when visiting Rome; her hotel bill, in London, was equal to her extra expenses.

Use the above information to calculate:
(a) the total cost, in francs, of the woman's visit to Paris;
(b) her extra expenses, in lire, in Rome;
(c) the amount, in pounds, of her hotel bill in London;
(d) her overall total cost, in dollars, in visiting Paris, Rome and London.
(Assume that the rates of exchange were: 4.10 francs to the dollar, in Paris; 850 lire to the dollar, in Rome; 2.40 dollars to the pound, in London.)

[AEB]

33 During the course of a year a motorist drove 28 000 km. For a quarter of this distance he drove in England and used a car with average petrol consumption of 14 litres per 100 km. For the remainder of this distance he drove in France and used a more economical car with an average petrol consumption of 8 litres per 100 km. Calculate:
(a) the total number of litres of petrol used;
(b) his average petrol consumption for the whole year giving the answer in litres per 100 km.
His total expenditure for the year on petrol was £630.

Given that the average price he paid for petrol in England was 20p per litre, calculate the total amount spent on petrol in France.

Assuming the average rate of exchange to be £1 = 8.4 francs, calculate the average price paid for petrol in France, giving the answer in francs per litre to the nearest tenth of a franc.

Calculate how much the motorist would have saved had he driven the more economical car in England as well as in France. [AEB]

3 Algebra

3.1 Directed Numbers

There is always great confusion between the concept of a directed number, and the operations of addition and subtraction. The author will try to make it clearer in this section by writing all numbers negative or positive like $(+2)$ or (-5).

(a) Addition

$$(+2) + (+5) = (+7)$$
$$(+5) + (-3) = (+2)$$
$$(-5) + (-7) = (-12)$$

> Addition of any negative number to any other number will always give a smaller answer.
>
> Addition of any positive number to any other number always gives a larger answer.

This is only helpful advice; the student still has to get the answer right!
The following summarizes the above more accurately:

> Addition of a positive number p increases the original number by p.
> Addition of a negative number n decreases the original number by n.

Note that in the third example above, if you decrease (-5) by 7 it becomes (-12).

(b) Subtraction

$$(+2) - (+5) = (-3)$$
$$(+5) - (-3) = (+8)$$
$$(-5) - (-7) = (+2)$$

> Subtraction of any positive number from any number always gives a smaller number.
>
> Subtraction of any negative number from any number always gives a larger number.

Subtraction of a positive number p decreases the original number by p.
Subtraction of a negative number n increases the original number by n.

Note that in the third subtraction example above, if you increase $\left(-5\right)$ by 7 you get $\left(+2\right)$.

(c) Multiplication and Division

$$\left(+5\right) \times \left(+4\right) = \left(+20\right) \qquad \left(-8\right) \times \left(+3\right) = \left(-24\right)$$

$$\left(+3\right) \times \left(-6\right) = \left(-18\right) \qquad \left(-4\right) \times \left(-5\right) = \left(+20\right)$$

$$\left(+8\right) \div \left(+2\right) = \left(+4\right) \qquad \left(-14\right) \div \left(+2\right) = \left(-7\right)$$

$$\left(+12\right) \div \left(-3\right) = \left(-4\right) \qquad \left(-40\right) \div \left(-5\right) = \left(+8\right)$$

The results here are easier to follow.

If two positive or negative numbers are multiplied or divided the answer is positive.
If a negative and a positive number are multiplied or divided, the answer is negative.

3.2   Commutative, Associative, Distributive, Brackets

If a, b, c are any numbers,

$$\left.\begin{array}{l} a + b = b + a \\ a \times b = b \times a \end{array}\right\} \text{commutative laws}$$

$$\left.\begin{array}{l} (a + b) + c = a + (b + c) \\ a \times (b \times c) = (a \times b) \times c \end{array}\right\} \text{associative laws}$$

$$\left.\begin{array}{l} a \times (b + c) = a \times b + a \times c \\ (a + b) \times c = a \times c + b \times c \end{array}\right\} \text{distributive laws}$$

Most of these rules will already be used quite competently by the student, but it is crucial to evaluate brackets correctly.

Worked Example 3.1

Evaluate without the use of a calculator:
(a) $(\frac{1}{2})^2 + (1 - \frac{2}{3})^2$,
(b) $(-\frac{2}{3} \times 1\frac{1}{2}) \div (\frac{1}{4} - (-\frac{1}{3})^2)$,
(c) $\frac{2}{3} \times (1\frac{1}{2} \times \{[\frac{2}{3} - (\frac{1}{2})^2] \div \frac{2}{5}\})$.

Solution

(a) $\frac{1}{4} + (\frac{1}{3})^2 = \frac{1}{4} + \frac{1}{9} = \frac{13}{36}$.
(b) $= (-\frac{2}{3} \times \frac{3}{2}) \div (\frac{1}{4} - \frac{1}{9}) = -1 \div \frac{5}{36}$
$= -1 \times \frac{36}{5} = -\frac{36}{5}$.
(c) Start with the inner brackets and work outwards:
$= \frac{2}{3} \times (\frac{3}{2} \times \{[\frac{2}{3} - \frac{1}{4}] \div \frac{2}{5}\})$
$= \frac{2}{3} \times (\frac{3}{2} \times \{\frac{5}{12} \div \frac{2}{5}\}) = \frac{2}{3} \times (\frac{3}{2} \times \frac{25}{24})$
$= \frac{2}{3} \times \frac{25}{16} = \frac{25}{24}$.

The following exercise is important; numbers have been chosen not to give too much heartache, and they should be evaluated without a calculator.

Check 12

1 $-8 - -2$	2 $-14 - -3 + -5$	3 $19 - -36 - 43$
4 $1\frac{3}{4} - 2\frac{7}{8}$	5 $1\frac{1}{12} - 1\frac{3}{4}$	6 $-8\frac{3}{8} - -3\frac{7}{8} + -2\frac{1}{2}$
7 $-\frac{3}{8} \div 2\frac{1}{2}$	8 $-\frac{1}{4} \times (\frac{1}{2} + -\frac{3}{8})$	9 $1\frac{3}{4} \times (-\frac{3}{8} - \frac{2}{5})$
10 $1\frac{1}{4} \times -1\frac{3}{4} \times -\frac{1}{2}$	11 $(-\frac{3}{4} \div -\frac{3}{5}) \div -\frac{1}{8}$	
12 $(1\frac{1}{2} + 1\frac{1}{3}) \div (1\frac{1}{2} - 1\frac{1}{7})$		13 $(\frac{5}{8} - \frac{7}{16}) \div [1 - (\frac{5}{8} \times \frac{4}{15})]$
14 $\{[(3\frac{1}{4} + 1\frac{2}{5}) \div 3\frac{1}{10}] - 1\frac{1}{2}\}$		15 $[(6\frac{1}{2})^2 \div 4\frac{1}{3}] \times (-\frac{3}{5} + 4\frac{1}{4})$

3.3 Substitution

One of the most common applications of algebra is in the use of formulae. There are many pitfalls in substituting into such formulae and the following examples illustrate techniques to avoid these where possible.

Worked Example 3.2

In the following question, $a = 2$, $b = -3$, $c = -\frac{1}{2}$. In each case find the value of x given by each formulae.
(a) $x = 2a + 3b + 4c$.
(b) $x = 3a - 5b$.
(c) $x = a^2 + b^2$.
(d) $x = 2b^2 + 4c^2$.
(e) $x = \dfrac{3ac}{b}$.
(f) $x = \dfrac{2}{a} + \dfrac{4}{c^2}$.
(g) $x = \sqrt{\dfrac{a + b}{c}}$.

Solution

(a) In full, $x = 2 \times 2 + 3 \times -3 + 4 \times -\frac{1}{2}$;
with brackets, $x = (2 \times 2) + (3 \times -3) + (4 \times -\frac{1}{2})$
$= 4 + -9 + -2 = -7$.

(b) In full, $x = 3 \times 2 - 5 \times -3$;
  with brackets, $x = (3 \times 2) - (5 \times -3)$
  $= 6 - -15 = 21$.

(c) In full, $x = 2^2 + -3^2$;
  with brackets, $x = (2)^2 + (-3)^2$
  $= 4 + 9 = 13$.

(d) In full, $x = 2 \times -3^2 + 4 \times -\frac{1}{2}^2$
  with brackets, $x = [2 \times (-3)^2] + [4 \times (-\frac{1}{2})^2]$
  $= [2 \times 9] + [4 \times \frac{1}{4}]$
  $= 18 + 1 = 19$.

(e) In full, $x = \dfrac{3 \times 2 \times -\frac{1}{2}}{-3}$;
  with brackets, $x = (3 \times 2 \times -\frac{1}{2}) \div -3$
  $= -3 \div -3 = 1$.

(f) In full, $x = \dfrac{2}{2} + \dfrac{4}{(-\frac{1}{2})^2}$;
  with brackets, $x = [2 \div 2] + [4 \div (-\frac{1}{2})^2]$
  $= 1 + [4 \div \frac{1}{4}]$
  $= 1 + 16 = 17$.

(g) In full, $x = \sqrt{\dfrac{2 + -3}{-\frac{1}{2}}}$;
  with brackets, $x = \sqrt{(2 + -3) \div -\frac{1}{2}} = \sqrt{-1 \div -\frac{1}{2}} = \sqrt{2}$.

You should try to see that liberal use of brackets makes the substitution much clearer to understand, and also makes it much easier to use a calculator, especially one which has a bracket facility.

3.4 Indices

In Chapter 2, simple problems with indices were developed using standard form. The rules of indices in general are as follows:

(1) $a^m \times a^n = a^{m+n}$ (5) $a^{q/p} = (\sqrt[p]{a})^q$ or $\sqrt[p]{a^q}$

(2) $a^m \div a^n = a^{m-n}$ (6) $a^0 = 1$

(3) $(a^m)^n = a^{mn}$ (7) $a^{-n} = \dfrac{1}{a^n}$

(4) $a^{1/p} = \sqrt[p]{a}$ (8) $(ab)^n = a^n b^n$

The use of these rules is shown in the following example.

Worked Example 3.3

Simplify the following:

(a) $(2^3)^2 \div 2^{-4}$ (b) $(49^{1/3})^{3/2}$ (c) $16^{-3/4}$ (d) $(a^2 b)^3 \div 4ab^3$

Solution

(a) $(2^3)^2 \div 2^{-4} = 2^6 \div 2^{-4}$ Rule 3
  $= 2^{10}$.   Rule 2

(b) $\quad (49^{1/3})^{3/2} = 49^{1/2} \qquad$ Rule 3

$\qquad\qquad\qquad\quad = 7. \qquad$ Rule 4

(c) $\quad 16^{-3/4} = \dfrac{1}{16^{3/4}} \qquad\qquad$ Rule 7

$\qquad\qquad = \dfrac{1}{(16^{1/4})^3} \qquad$ Rule 5

$\qquad\qquad = \dfrac{1}{2^3} \qquad\qquad$ Rule 4

$\qquad\qquad = \dfrac{1}{8}.$

(d) $\quad (a^2 b)^3 \div 4ab^3 = a^6 b^3 \div 4ab^3 \qquad$ Rule 8

$\qquad\qquad\qquad = \dfrac{a^6 b^3}{4ab^3} = \dfrac{a^5 b^0}{4} \qquad$ Rule 2

$\qquad\qquad\qquad = \dfrac{a^5}{4} \qquad\qquad$ Rule 6.

A useful trick when working with indices is always to get rid of negative powers first if possible.

Now try the following:

Check 13

Simplify the following:

1 $\sqrt{10^4}$	2 $\sqrt[3]{27^2}$	3 $(0.09)^{1/2}$	4 $125^{-2/3}$
5 $27^{-1/3}$	6 $(2^{1/5})^{10}$	7 $(3^{2/3})^0$	8 $49^{1/2}$
9 $32^{2/5}$	10 $625^{-1/2}$	11 $4x^5 \div 2x^2$	12 $5p^{-3} \div 4p^5$
13 $3x^{-2} \times 4x^{-4}$	14 $(-3m)^3$	15 $\sqrt[3]{27x^6}$	16 $\sqrt{25y^{16}}$
17 $(8q^2)^{2/3}$	18 $\left(\dfrac{1}{u^{-2}}\right)^{1/2}$	19 $\left(-\dfrac{1}{3v^2}\right)^{-2}$	20 $1 \div (\tfrac{4}{9})^{-3/2}$

3.5 Like and Unlike Terms

The following examples indicate which terms in algebra are *like*, and which are *unlike*:

$4x + 7x = 11x.$
$2x^2 + 3x^2 = 5x^2.$

$4xy + 7x^2 y$ cannot be simplified (*unlike* terms).

$3xy + 7x^2 - 2xy = xy + 7x^2$ (first and last terms are *like*).

In an expression such as $11xy$, the number 11 is called the *coefficient* of xy.

3.6 Removal of Brackets

Worked Example 3.4

Remove brackets from the following expressions, and simplify answers where possible.

(a) $4x(a + x) + 3a(x + a)$
$\qquad = 4xa + 4x^2 + 3ax + 3a^2$ (using distributive law and indices rules)
$\qquad = 4x^2 + 7ax + 3a^2$.
    (*Note*: $4xa$ and $3ax$ are *like* terms.)
(b) $2y(x - y) - 3x(y - x)$
                                   — Subtraction of a negative quantity is the
$\qquad = 2yx - 2y^2 - 3xy \oplus 3x^2$.   same as addition; see Chapter 2.
(c) $(2x - 1)(3x + 1)$
$\qquad = (2x \times 3x) + (2x \times 1) - (1 \times 3x) - (1 \times 1)$
$\qquad = 6x^2 + 2x - 3x - 1 = 6x^2 - x - 1$.

3.7   Factors

The reverse process of putting brackets back into an algebraic expression is called *factorizing*.

(a) Two Terms

Basically this is the reverse of the distributive law:
(i) $8x + 16y = 8(x + 2y)$.
(ii) $4p^2 + 8pq = 4p(p + 2q)$. $\begin{bmatrix} p(4p + 8q) \text{ is correct but not completely} \\ \text{factorized.} \end{bmatrix}$

(b) Four Terms

(i) $6a + 12 + 4ab + 8b$
$\qquad = 6(a + 2) + 4b(a + 2) = (6 + 4b)(a + 2)$
$\qquad\qquad\qquad\qquad\qquad = 2(3 + 2b)(a + 2)$.
(ii) $a^2 - 3a - ab + 3b$
                           — $\begin{bmatrix} \text{Note that this is negative, because} \\ -b \times -3 = +3b. \end{bmatrix}$
$\qquad = a(a - 3) - b(a - 3)$
$\qquad = (a - b)(a - 3)$.

(c) Three Terms (Quadratic)

In Worked Example 3.4(c) we showed that
$$(2x - 1)(3x + 1) = 6x^2 - x - 1;$$
the expression $6x^2 - x - 1$ is called a *quadratic* expression.
     The process of factorizing a quadratic is not easy and will be considered in stages.
(i) $x^2 + 5x + 6 \quad = x^2 + 2x + 3x + 6$
$\qquad\qquad\qquad\qquad = x(x + 2) + 3(x + 2)$
$\qquad\qquad\qquad\qquad = (x + 3)(x + 2)$.
(ii) $x^2 + x - 6 \quad = x^2 + 3x - 2x - 6$
$\qquad\qquad\qquad\qquad = x(x + 3) - 2(x + 3)$
$\qquad\qquad\qquad\qquad = (x - 2)(x + 3)$.
(iii) $x^2 - 7x + 10 = x^2 - 5x - 2x + 10$
$\qquad\qquad\qquad\qquad = x(x - 5) - 2(x - 5)$
$\qquad\qquad\qquad\qquad = (x - 2)(x - 5)$.

The process is really trial and error, and practice makes perfect.

(iv) $2x^2 - x - 6 = 2x^2 + 3x - 4x - 6$
$$= x(2x + 3) - 2(2x + 3)$$
$$= (x - 2)(2x + 3).$$

(v) $18x^2 + 45x - 8 = 18x^2 + 48x - 3x - 8$
$$= 6x(3x + 8) - (3x + 8)$$
$$= (6x - 1)(3x + 8).$$

When splitting up into four terms, notice that the product of the two middle terms equals the product of the two outside terms, e.g.:

(i) $x^2 \times 6 = 2x \times 3x = 6x^2$,

(iii) $x^2 \times 10 = -5x \times -2x = 10x^2$,

(v) $18x^2 \times -8 = 48x \times -3x = -144x^2$.

It is not always possible to put a quadratic expression into brackets; the following rule can save you a lot of time.

For the expression $ax^2 + bx + c$, work out $b^2 - 4ac$.

If it is a perfect square it factorizes, if not it doesn't.

For example:

(i) $x^2 + 6x + 8$ $a = 1, b = 6, c = 8$;
$b^2 - 4ac = 36 - 32 = 4$, which *is* a perfect square.
In fact $x^2 + 6x + 8 = (x + 4)(x + 2)$.

(ii) $x^2 - 4x - 12$ $a = 1, b = -4, c = -12$;
$b^2 - 4ac = 16 + 48 = 64$, which *is* a perfect square.

The alternative method which also involves trial and error is illustrated as follows:

Factorize $6x^2 + 19x - 20$.

Consider this   Consider this
number first.   number second.

$6x^2$ suggests $3x \times 2x$.
-20 suggests 5×-4 or 2×-10.
Try $(3x + 5)(2x - 4) = 6x^2 + 10x - 12x - 20$
$$= 6x^2 - 2x - 20. \qquad \times$$
Try $(2x - 10)(3x + 2) = 6x^2 + 4x - 30x - 20$
$$= 6x^2 - 26x - 20. \qquad \times$$
There are many possibilities.
In fact $(6x - 5)(x + 4) = 6x^2 - 5x + 24x - 20$
$$= 6x^2 + 19x - 20. \qquad \checkmark$$

(d) $a^2 - b^2$

This expression is known as a *difference of two squares*.
$a^2 - b^2 = (a - b)(a + b)$.
Example: $16x^2 - 25y^2 = (4x)^2 - (5y)^2 = (4x - 5y)(4x + 5y)$.

(e) $a^2 \pm 2ab + b^2$

This expression is a *perfect square*.
$a^2 + 2ab + b^2 = (a + b)(a + b) = (a + b)^2$.
$a^2 - 2ab + b^2 = (a - b)^2$.

Check 14

Factorize if possible the following:

1 $ay + 4y^2$	2 $3t^3 + 2t^2 + 5t$	3 $p^2q^3 - q^2p^3$
4 $7(p - 2) - 3(p - 2)$	5 $p(y - z) - q(z - y)$	6 $qr + rs - ps - pq$
7 $t^2 - 3t - tp + 3p$	8 $ax + bx + cx - ay - by - cy$	
9 $x^2 - 11x + 24$	10 $x^2 + 2x - 3$	11 $x^2 + 11x - 26$
12 $x^2 - 7x + 8$	13 $2x^2 + 13x + 15$	14 $4x^2 + 8x + 3$
15 $6x^2 - 5x + 1$	16 $9x^2 - 7x - 2$	17 $4x^4 - 3x^2 - 1$
18 $4x^2 - 25$	19 $1 - x^2$	20 $\pi R^2 - \pi r^2$
21 $(x - 3)^2 - 9$	22 $121x^2y^2 - 4$	23 $4x^2 - 12xy + 9y^2$
24 $25x^2 - 10xy + y^2$	25 $25 - 10b + b^2$	

3.8   LCM and HCF

The lowest common multiple of (LCM) of two numbers is the smallest number that both numbers divide into exactly. It can be found by first expressing each number as the product of its prime factors. Consider the following example.

To find the LCM of 48 and 84:

```
2)48          2)84
2)24          2)42
2)12          3)21
2) 6          7) 7
3) 3             1
   1
```

$\therefore$ $48 = 2^4 \times 3$. $\therefore 84 = 2^2 \times 3 \times 7$.

Take the highest power of every prime number occurring in the two products and multiply these powers together. Hence,

$$\text{LCM} = 2^4 \times 3 \times 7 = 336.$$

To find the HCF of these numbers, take the lowest power of each prime number occurring in both numbers, and multiply them together. Hence,

$$\text{HCF} = 2^2 \times 3 = 12.$$

3.9   LCM and HCF in Algebra

The method is slightly easier in algebra, because most of the work is done already. For example, the LCM of $4a^3bc$ and $6a^2b^4c^2$ is

$$12 \times a^3 \times b^4 \times c^2 = 12a^3b^4c^2.$$

The HCF is $2 \times a^2 \times b \times c = 2a^2bc$.

3.10 Algebraic Fractions

(a) Addition or Subtraction

$$\frac{4}{3ab} + \frac{1}{4a^2 b} = \frac{4 \times 4a}{12a^2 b} + \frac{1 \times 3}{12a^2 b} = \frac{16a + 3}{12a^2 b}$$

$$\uparrow$$

LCM of $3ab$ and $4a^2 b$

(b) Multiplication

$$\frac{3ab}{4c} \times \frac{6a^2 c}{5bt} \text{ in full } = \frac{3 \times a \times \cancel{b}}{\cancel{4} \times c} \times \frac{\overset{3}{\cancel{6}} \times a \times a \times \cancel{c}}{5 \times \cancel{b} \times t}$$

$$\underset{2}{}$$

$$= \frac{9a^3}{10t} .$$

(c) Division

$$\frac{4xy^2}{9t} \div \frac{8x^2 y^3}{7t^2} = \frac{4xy^2}{9t} \times \frac{7t^2}{8x^2 y^3}$$

$$= \frac{\overset{1}{\cancel{4}} \times \cancel{x} \times \cancel{y} \times \cancel{y}}{9 \times \cancel{t}} \times \frac{7 \times \cancel{t} \times t}{\underset{2}{\cancel{8}} \times \cancel{x} \times x \times \cancel{y} \times \cancel{y} \times y}$$

$$= \frac{7t}{18xy} .$$

With practice, the 'in full' stage can be omitted.

Check 15

Evaluate and simplify if possible the following:

1 (a) $\dfrac{1}{x} + \dfrac{1}{x^2}$ (b) $\dfrac{2}{a} + \dfrac{3}{ab}$ (c) $\dfrac{4}{t} - \dfrac{3}{2t^2}$

(d) $\dfrac{x}{2} + \dfrac{x}{3} + \dfrac{x}{4}$ (e) $\dfrac{4}{y} - \dfrac{2}{3y} + \dfrac{5}{y}$ (f) $\dfrac{4x}{3y} - \dfrac{5y}{7x}$

(g) $4x + \dfrac{1}{x}$ (h) $\dfrac{2a + 5b}{4} + \dfrac{3a + 2b}{8}$ (i) $\dfrac{x + 2}{3} - \dfrac{5}{6}$

(j) $\dfrac{4}{(x - 1)} + \dfrac{3}{(x + 2)}$ (k) $\dfrac{1}{x} - \dfrac{1}{(x - 1)}$ (l) $\dfrac{1}{x^2 - 5x + 6} + \dfrac{x}{(x - 3)}$

(m) $\dfrac{x + 1}{x - 1} - \dfrac{x - 1}{x + 1}$

2 (a) $\dfrac{x}{y^2} \times \dfrac{2y}{x}$ (b) $\dfrac{3pq}{t} \times \dfrac{4t^2}{9p}$ (c) $\dfrac{5t}{6q} \times \dfrac{18q^2}{25t^2}$

(d) $\dfrac{x}{y} \times \dfrac{2y}{3x} \times \dfrac{4x}{5y}$ (e) $\dfrac{p^2}{q} \times \dfrac{2q^2}{3p} \times \dfrac{4}{5pq}$ (f) $\dfrac{pq}{r} \div \dfrac{3p}{2r}$

(g) $\dfrac{a-b}{c} \times \dfrac{c^2}{a^2-b^2}$ (h) $\dfrac{5pq}{6t^2} \div \dfrac{3pq}{5t}$ (i) $\dfrac{16t^2}{5p} \div \dfrac{4t}{5p^2}$

(j) $\dfrac{a+b}{c} \div \dfrac{a-b}{c}$

3.11   Forming Expressions

The ability to write down an expression in terms of algebraic quantities is extremely important if problem solving is to be attempted. The following examples should illustrate the necessary techniques that need to be mastered.

(a) A number x is multiplied by 4, and 3 is subtracted from the answer. The result is divided by 4. What is the result?

Answer $= \dfrac{(4x-3)}{4}$.   Brackets are important in order to indicate that all of $4x - 3$ is divided by 4.

(b) Find the cost of p apples at t pence a dozen, and q pears at 6 pence each.

The cost of one apple is $\dfrac{t}{12}$ pence.

Hence the cost of p apples is $\dfrac{t}{12} \times p = \dfrac{tp}{12}$.

The cost of the pears is $6q$ pence.

The total cost is $\dfrac{tp}{12} + 6q$.

(c) A car travels at x m.p.h. for 30 min, and then travels a further k miles for $\frac{1}{4}$ hour. Find the average speed for the journey.

Since   Average speed $= \dfrac{\text{Total distance}}{\text{Total time}}$,

We need to find the distance travelled in the first 30 min, i.e. in $\frac{1}{2}$ hour.

Distance = Speed × Time,
$\therefore$ distance $= \frac{1}{2}x$
$\therefore$ total distance $= \frac{1}{2}x + k$
     total time $= \frac{1}{2} + \frac{1}{4} = \frac{3}{4}$.
$\therefore$ Average speed $= \dfrac{\frac{1}{2}x + k}{\frac{3}{4}} = \frac{4}{3}(\frac{1}{2}x + k)$.

Answers should always be simplified as much as possible.

Check 16

Find in as simple a form as possible expressions for the following.

1   A number x is doubled, 5 is subtracted from this and the answer is divided in the ratio 2 : 1. What is the smallest part?

2   What is the total cost of p pencils at y pence for 20 and q pens at £k each. Give your answer in pounds.

3   What is the area of a rectangle in cm^2, if it measures x mm by y cm?

4 How many minutes are there between y min to 8.00 hours and x min after 10.30 hours?

5 A labourer earned £t a week for y weeks of the year, and £v a week for the remainder except for 2 weeks when he had no income. How much did he earn in the year?

6 I could buy p pens for £x. If the price is increased by 10 per cent, how many pens can I buy for the same amount?

7 A batsman scored f fours, s sixes and 9 singles in an innings. How many runs did he score?

8 A rectangle measures p cm by q cm. ($p < q$). If the shorter side is increased by 10 per cent, and the longer side is decreased by 8 per cent, find an expression for the percentage change in its area.

9 If income tax is charged at I pence in the pound, how much tax is paid by a person who earns £E, but is allowed deductions of £D before tax is calculated? Give your answer in pounds.

10 Telephone cable weighing W kg/m is wound on to a reel which weighs M kg. If the reel can hold 85 m, find the total weight of the reel and cable.

3.12   Linear Equations

You should be able to solve simple equations. The following examples, however, should refresh your memory and indicate simple procedures.

Remember, any number can be added or subtracted from either side of an equation without altering it. Similarly, either side of an equation can be multiplied or divided by any number without altering it.

(a) $7x - 5 = 4x + 11$;

add 5 to each side: $\qquad 7x = 4x + 16$;

take $4x$ from each side: $\quad 3x = 16$;

divide each side by 3: $\qquad x = \dfrac{16}{3}$.

(b) $4 - (3x + 1) = 2(x - 5)$;

remove brackets first: $\qquad 4 - 3x - 1 = 2x - 10$;

collect terms on the left: $\qquad 3 - 3x = 2x - 10$;

add 10 to each side: $\qquad 13 - 3x = 2x$;

add $3x$ to each side: $\qquad 13 = 5x$;

divide each side by 5: $\qquad \dfrac{13}{5} = x$.

(c) $\dfrac{2x + 1}{4} - \dfrac{3x - 5}{2} = \dfrac{x}{3}$;

to avoid problems with signs, rewrite the equation with brackets inserted; i.e.

$$\frac{(2x + 1)}{4} - \frac{(3x - 5)}{2} = \frac{x}{3};$$

multiply the equation by the common denominator, i.e. the LCM of 2, 3, 4, which is 12:

$$\frac{\overset{3}{\cancel{12}}(2x + 1)}{\cancel{4}} - \frac{\overset{6}{\cancel{12}}(3x - 5)}{\cancel{2}} = \overset{4}{\cancel{12}} \times \frac{x}{\cancel{3}}; \qquad \text{cancel;}$$

remove brackets: $6x + 3 - 18x + 30 = 4x$;
collect terms: $-12x + 33 = 4x$;
add $12x$ to each side: $33 = 16x$;

divide each side by 16: $\dfrac{33}{16} = x$.

(d) $\dfrac{7}{(x-1)} = \dfrac{4}{(x+1)}$;

the common denominator is $(x-1)(x+1)$;
multiply the equation by this:

$$\cancel{(x-1)}(x+1) \times \frac{7}{\cancel{(x-1)}} = \frac{4}{\cancel{(x+1)}} \times (x-1)\cancel{(x+1)}; \qquad \text{cancel;}$$

$\therefore \quad 7(x+1) = 4(x-1),$
$\qquad 7x + 7 = 4x - 4,$
$\qquad 3x = -11,$
$\therefore \qquad x = -\dfrac{11}{3}$.

3.13   Quadratic Equations

(a) By Factors

Consider $x^2 - 5x + 6 = 0$. In section 3.7(c) we factorized quadratic expressions.
  Hence $(x-3)(x-2) = 0$,
$\therefore$ either $x - 3 = 0$, or $x - 2 = 0$.
We get two solutions: $x = 3$ or $x = 2$.

A further example: $6x^2 - x - 2 = 0$;
factorizing, $(2x + 1)(3x - 2) = 0$
$\therefore 2x + 1 = 0 \quad$ or $\quad 3x - 2 = 0$,
$\qquad 2x = -1 \quad$ or $\quad 3x = 2$,
$\therefore \qquad x = -\tfrac{1}{2} \quad$ or $\quad x = \tfrac{2}{3}$.

(b) By Formula

The equation $x^2 + x - 1 = 0$ cannot be solved by factors, because it does not factorize. The following formula can be used. A proof is given.
  Consider the equation $ax^2 + bx + c = 0$.

Divide by a: $x^2 + \dfrac{bx}{a} + \dfrac{c}{a} = 0$.

Add $\dfrac{b^2}{4a^2}$ to both sides. This is part of the technique of completing the square.

$$\therefore x^2 + \frac{bx}{a} + \frac{b^2}{4a^2} + \frac{c}{a} = \frac{b^2}{4a^2} ,$$

$$\left(x + \frac{b}{2a}\right)^2 = \frac{b^2}{4a^2} - \frac{c}{a} \qquad \text{Check by squaring out the bracket.}$$

$$= \frac{b^2 - 4ac}{4a^2} .$$

Take the square root of each side, remembering that it could be positive or negative:

$$x + \frac{b}{2a} = \pm \frac{\sqrt{b^2 - 4ac}}{2a}.$$

Subtract $\frac{b}{2a}$ from both sides:

$$\boxed{x = \frac{-b \pm \sqrt{b^2 - 4ac}}{2a}.}$$

In the equation $x^2 + x - 1 = 0$, $a = 1$, $b = 1$, $c = -1$.

$$\therefore x = \frac{-1 \pm \sqrt{1 + 4}}{2} = \frac{-1 \pm \sqrt{5}}{2}.$$

Hence, $x = 0.618$ or -1.618.

3.14   Simultaneous Equations (Linear)

Linear simultaneous equations can be solved by elimination, as shown here, by graphs, see section 5.9, or by matrices, see section 9.8. See also section 3.19 for non-linear equations.

(a) Solve

$$3x + 4y = 2 \tag{1}$$
$$5x + 3y = 1. \tag{2}$$

Multiply (1) by 5: $15x + 20y = 10$ (3)
multiply (2) by 3: $15x + 9y = 3.$ (4)

Each equation has the same number of x's. The equations can be subtracted:

$$(3) - (4): \quad 11y = 7, \quad \therefore y = \tfrac{7}{11}.$$

Substitute this value into any of the equations, i.e. (1):

$$3x + \tfrac{28}{11} = 2, \quad \therefore 3x = 2 - \tfrac{28}{11} = -\tfrac{6}{11}, \quad \therefore x = -\tfrac{2}{11}.$$

The solution to the equations is $x = -\tfrac{2}{11}$, $y = \tfrac{7}{11}$.

(b) A slightly harder example:

$$5x - 4y = 14 \tag{1}$$
$$7x + 9y = 18. \tag{2}$$

Multiply (1) by 9: $45x - 36y = 126$ (3)
multiply (2) by 4: $28x + 36y = 72.$ (4)

This time, y can be eliminated by adding the equations:

$$(3) + (4): \quad 73x = 198, \quad \therefore x = \tfrac{198}{73} = 2.71 \ (2 \text{ d.p.}).$$

Substitute in (2):

$$18.97 + 9y = 18, \quad \therefore 9y = -0.97, \quad \therefore y = -0.11.$$

The solution is $x = 2.71$, $y = -0.11$.

Notes:

(1) It may be possible to eliminate x or y without multiplying either equation (see the following example).
(2) It doesn't matter which variable is eliminated.

(c) Solve

$$4x = 5 + 3y, \tag{1}$$
$$3y - 5x = -4. \tag{2}$$

Before solving any simultaneous equations, the variables must be on the same side of the equation in the same order.

Rewrite (1): $\qquad\qquad 4x - 3y = 5.$ $\qquad\qquad$ (3)
Rewrite (2): $\qquad\qquad -5x + 3y = -4.$ $\qquad\qquad$ (4)

(3) + (4): $\quad -x = 1, \qquad \therefore x = -1.$
Substitute in (4): $\quad 5 + 3y = -4$
$\qquad\qquad\qquad\qquad 3y = -9, \qquad \therefore y = -3.$
The solution is $x = -1$, $y = -3$.

Now practise solving the following equations:

Check 17

1 Solve for x:
 (a) $2x + 3 = 7$; $\qquad\qquad\qquad$ (b) $8 - x = 4$;
 (c) $4x - 1 = 7x + 3$; $\qquad\qquad$ (d) $4(x + 2) = 3 + (x - 5)$;
 (e) $2(x + 1) + 3(x - 2) = 15$; $\qquad$ (f) $\frac{1}{2}x + 4 = 3x$;
 (g) $2(x + 3) = 9$; $\qquad\qquad\quad$ (h) $6 - 5(x + 1) = 9$;
 (i) $\dfrac{x}{5} + \dfrac{2x}{3} = \dfrac{1}{2}$ $\qquad\qquad$ (j) $\dfrac{x + 1}{2} + \dfrac{x - 1}{2} = \dfrac{1}{2}$;
 (k) $\dfrac{2x - 1}{5} + \dfrac{3x - 2}{4} = 1$; $\qquad$ (l) $\dfrac{(x + 2)}{5} - \dfrac{3x}{5} = 4$;
 (m) $\dfrac{7}{x} = 9$; $\qquad\qquad\qquad$ (n) $\dfrac{2}{(x - 3)} = \dfrac{4}{(x + 2)}$;
 (o) $\dfrac{1}{(x + 2)(x + 5)} = \dfrac{1}{(x + 3)(x - 1)}$.

2 Solve where possible the following quadratic equations:
 (a) $x^2 - 2x - 3 = 0$; $\qquad\qquad$ (b) $x^2 - 10x + 16 = 0$;
 (c) $2x^2 + x - 1 = 0$; $\qquad\qquad$ (d) $3x^2 - 10x + 3 = 0$;
 (e) $x^2 - 2x + 1 = 0$; $\qquad\qquad$ (f) $x^2 - 9x = 0$;
 (g) $x^2 + 3x - 9 = 0$; $\qquad\qquad$ (h) $x^2 + 2x - 1 = 0$;
 (i) $x^2 + 8x = 3$; $\qquad\qquad\quad$ (j) $4x^2 = 9x + 1$;
 (k) $x + \dfrac{1}{3x} = 2$; $\qquad\qquad$ (l) $\dfrac{x + 2}{x + 5} = \dfrac{2x + 3}{x - 1}$;
 (m) $4x^2 + x + 7 = 0$; $\qquad\qquad$ (n) $\dfrac{1}{x - 1} - \dfrac{1}{x + 1} = 6$;
 (o) $\dfrac{1}{x} + x = 2$.

3 Solve the following simultaneous equations:
 (a) $x + y = 7$, $2x - y = 8$; $\qquad$ (b) $3x - 2y = 4$, $5x - 2y = 0$;
 (c) $4x - 3y = 5$, $7x - 5y = 9$; $\qquad$ (d) $2x + y = 8$, $5x - y = 6$;
 (e) $x + 3y = 2$, $2x - 4y = 1$; $\qquad$ (f) $3x + 2y = 13$, $2x + 3y = 12$;
 (g) $4x - 5y = 21$, $6x + 7y = -12$; $\quad$ (h) $3x + 2y = 17$, $4y - x = -8$;
 (i) $3y = 2x - 7$, $3x = 13 + 2y$; $\qquad$ (j) $2x = 3y + 5$, $5x = 3y - 4$.

3.15 Forming Equations

The techniques of making algebraic expressions and solving equations can now be combined to solve simple problems.

Worked Example 3.5

The sum of £61 is divided between Arthur, Brenda and Catherine. Arthur has twice as much as Brenda, and Catherine has £5 less than Arthur and Brenda together. How much do all three have each?

Solution

Use a suitable letter to denote one unknown quantity.

Let Brenda have £M, then Arthur has £$2M$. Catherine has £$(M + 2M) - 5 = £3M - 5$.

The total is £61, $\therefore M + 2M + 3M - 5 = 61$,

                          $\therefore 6M = 66$, $\therefore M = 11$.

$\therefore$ Arthur has £22, Brenda has £11 and Catherine £28.

Worked Example 3.6

A machine produces two types of bolt. Bolt A is produced at the rate of x per minute, and bolt B is produced at the rate of $(x - 6)$ per minute.
(a) Write down an expression in seconds for the time taken to produce each bolt.
(b) If it takes $1\frac{1}{6}$ seconds longer to produce bolt B than bolt A, write down an equation in x and solve it.

Solution

(a) If x of bolt A are produced in 60 seconds, then one is produced in $\dfrac{60}{x}$ seconds. Similarly, one of B is produced in $\dfrac{60}{x - 6}$ seconds.

(b) Seeing that the difference is $1\frac{1}{6}$ seconds, and $\dfrac{60}{x - 6}$ is larger than $\dfrac{60}{x}$ (because less bolts are produced per minute), it follows that:

$$\frac{60}{x - 6} = \frac{60}{x} + \frac{7}{6};$$

multiply by common denominator $6x(x - 1)$:

$$\therefore 360x = 360(x - 6) + 7x(x - 6)$$
$$\therefore 360x = 360x - 2160 + 7x^2 - 42x$$
$$\therefore 7x^2 - 42x - 2160 = 0$$
$$\therefore (7x + 90)(x - 24) = 0.$$

x cannot be $-\dfrac{90}{7}$, $\therefore x = 24$.

Worked Example 3.7

A bill of £355 was paid using £5 notes and £20 notes. If 35 notes were used altogether, find how many of each were used.

Solution

It is not possible to solve this problem with just one unknown.
Let the number of £5 notes be f.
Let the number of £20 notes be t.
For the money: $5f + 20t = 355.$ (1)
For the notes: $f + t = 35.$ (2)
Divide (1) by 5: $f + 4t = 71.$ (3)
        (3) − (2): $3t = 36,$ $\therefore t = 12,$ $f = 23.$
Thus 23 £5 notes and 12 £20 notes were used.

3.16   Changing the Subject of Formulae

In a short space this is best illustrated by a number of examples.

Worked Example 3.8

Make x the subject of the following formulae:
(a) $t = ax + b$ (b) $t = a(x + b);$ (c) $t = 2a\sqrt{x + y};$

(d) $y = \dfrac{1}{t}(ax - k);$ (e) $y = \dfrac{2t}{(1 + x)};$ (f) $p = a\sqrt{x} + \dfrac{q}{t};$

(g) $y = \dfrac{ax + b}{cx + d};$ (h) $y = \dfrac{t}{1 - x^2}.$

Solution

It is important to realise that the following solutions are not the only ones, but are an attempt at a systematic approach. Short cuts are often possible with experience.

(a) $t = ax + b.$

   Subtract b from each side: $t - b = ax.$

   $\div a$: $\dfrac{t - b}{a} = x,$

   or, better, $x = \dfrac{(t - b)}{a}.$

(b) $t = a(x + b).$

   Remove brackets: $t = ax + ab$

   Subtract ab from each side: $t - ab = ax.$

   $\div a$: $x = \dfrac{(t - ab)}{a}.$

(c) $t = 2a\sqrt{x + y}.$
   Square both sides (remember to square $2a$):

$$t^2 = 4a^2(x + y)$$

   Remove brackets: $t^2 = 4a^2 x + 4a^2 y.$

Subtract $4a^2y$ from each side:
$$t^2 - 4a^2y = 4a^2x.$$

$\div 4a^2$:
$$x = \frac{t^2 - 4a^2y}{4a^2}.$$

(d)
$$y = \frac{1}{t}(ax - k).$$

Remove brackets:
$$y = \frac{ax}{t} - \frac{k}{t}.$$

$\times t$:
$$yt = ax - k.$$

$+ k$ each side:
$$yt + k = ax.$$

$\div a$:
$$x = \frac{(yt + k)}{a}.$$

(e)
$$y = \frac{2t}{1 + x}.$$

$\times$ common denominator $(1 + x)$:
$$y(1 + x) = 2t.$$

Remove brackets:
$$y + xy = 2t.$$

$- y$ each side:
$$xy = 2t - y.$$

$\div y$:
$$x = \frac{(2t - y)}{y}.$$

(f)
$$p = a\sqrt{x} + \frac{q}{t}.$$

$- \dfrac{q}{t}$ each side:
$$p - \frac{q}{t} = a\sqrt{x}.$$

Square both sides:
$$\left(p - \frac{q}{t}\right)^2 = a^2x.$$

Remove brackets:
$$p^2 + \frac{q^2}{t^2} - \frac{2pq}{t} = a^2x$$

$\times$ common denominator t^2:
$$p^2t^2 + q^2 - 2pqt = a^2t^2x$$

$\div a^2t^2$:
$$x = \frac{p^2t^2 + q^2 - 2pqt}{a^2t^2}.$$

(g)
$$y = \frac{ax + b}{cx + d}$$

Common denominator $cx + d$,
$$\therefore \quad y(cx + d) = ax + b.$$

Remove brackets:
$$ycx + yd = ax + b,$$
$$\therefore \quad ycx - ax = b - yd.$$

Factorize side containing new subject x:
$$x(yc - a) = b - yd.$$

$\div (yc - a)$:
$$x = \frac{b - yd}{yc - a}.$$

(h)
$$y = \frac{t}{1 - x^2}.$$

Common denominator $(1 - x^2)$,

$$\therefore \quad y(1 - x^2) = t.$$

Remove brackets. $\qquad y - yx^2 = t,.$

$$\therefore \quad y - t = yx^2$$

$\div y$: $\qquad\qquad\qquad x^2 = \dfrac{y - t}{y}.$

Take square root: $\quad x = \sqrt{\dfrac{y - t}{y}} \qquad$ (remember x could be negative).

3.17 Remainder Theorem

The remainder theorem states that if a polynomial function $f(x)$ (e.g. $x^3 + 2x + 1$, $3x^2 - 5x + 6$) is divided by $(x - a)$, the remainder is $f(a)$. Hence, if $f(a) = 0$, $(x - a)$ must be a factor of the original expression. This is extremely useful in factorizing higher-order polynomials than quadratics.

Worked Example 3.9

Factorize
(a) (i) $x^3 + x^2 - 10x + 8$,
   (ii) $2x^3 + 7x^2 + 2x - 3$.

Solution

(a) Let $\qquad\qquad f(x) = x^3 + x^2 - 10x + 8.$

By trial, $\qquad\qquad f(1) = 1 + 1 - 10 + 8 = 0.$

$\therefore (x - 1)$ is a factor.

By division,

$$
\begin{array}{r}
x^2 + 2x - 8 \\
x - 1 \overline{\smash{)}\ x^3 + x^2 - 10x + 8} \\
\underline{x^3 - x^2} \\
2x^2 - 10x \\
\underline{2x^2 - 2x} \\
-8x + 8 \\
\underline{-8x + 8} \\
0 \longleftarrow \text{no remainder}
\end{array}
$$

$\therefore f(x) = (x - 1)(x^2 + 2x - 8)$
$\qquad = (x - 1)(x - 2)(x + 4).$

(b) Let $\qquad\qquad f(x) = 2x^3 + 7x^2 + 2x - 3.$

This is less obvious:

$$f(1) = 2 + 7 + 2 - 3 = 8.$$
$$f(2) = 16 + 28 + 4 - 3 = 45.$$

This doesn't look too hopeful; try negative values:

$$f(-1) = -2 + 7 - 2 - 3 = 0.$$

$\therefore x + 1$ is a factor.

By division,

$$\begin{array}{r} 2x^2 + 5x - 3 \\ x + 1 \overline{\smash{\big)}\, 2x^3 + 7x^2 + 2x - 3} \\ \underline{2x^3 + 2x^2} \\ 5x^2 + 2x \\ \underline{5x^2 + 5x} \\ -3x - 3 \\ \underline{-3x - 3} \\ 0 \end{array}$$

$$\begin{aligned} f(x) &= (x + 1)(2x^2 + 5x - 3) \\ &= (x + 1)(2x - 1)(x + 3). \end{aligned}$$

Worked Example 3.10

The cubic function $x^3 + ax^2 + 7x + b$ leaves a remainder 2 when divided by $(x - 3)$ and a remainder 1 when divided by $x + 2$. Find a and b.

Solution

The remainder theorem can be used without having to do any division. The remainder when divided by $(x - 3)$ is given by $f(3)$. Since this remainder is 2, it follows that:

$$f(3) = 2,$$

i.e. $\quad 3^3 + 9a + 21 + b = 2,$

$$\therefore \quad 9a + b = -46. \tag{1}$$

Similarly, $\qquad\qquad\qquad f(-2) = 1,$

$$\therefore \quad (-2)^3 + 4a - 14 + b = 1,$$

$$\therefore \quad 4a + b = 23. \tag{2}$$

(1) − (2): $\qquad\qquad\qquad 5a = -69.$

$$\therefore \quad a = -13.8, \; b = 78.2.$$

3.18   Variation and Proportion

In section 2.8, the idea of two quantities being proportional was introduced. The phrase 'varies (as)' can also be used. In general, if A varies as or is proportional to B, then $A = kB$, where k is the constant of proportionality. Other types of variation are possible.

If y is inversely proportional to x, then $y = \dfrac{k}{x}$.

If y varies as x^2, $\qquad\qquad y = kx^2$.
(y is proportional to x^2.)

This means that if x^2 is doubled, so is y. In order to double x^2, x would have to be multiplied by a factor $\sqrt{2}$.

A could be related to more than one quantity. See Worked Example 3.13.

Worked Example 3.11

If y is proportional to x^2, and $y = 4$ when $x = 8$, find y when $x = 2$.

Solution

Let $\qquad\qquad\qquad\qquad\qquad\qquad y = kx^2.$

$y = 4, x = 8, \qquad\qquad\qquad\qquad \therefore\quad 4 = k \times 8^2,$

$\qquad\qquad\qquad\qquad\qquad\qquad \therefore\quad k = \tfrac{4}{64} = \tfrac{1}{16},$

$\qquad\qquad\qquad\qquad\qquad\qquad \therefore\quad y = \tfrac{1}{16}x^2.$

$\therefore\quad$ If $x = 2, y = \tfrac{1}{16} \times 4 = \tfrac{1}{4}.$

3.19   Simultaneous Equations, One Linear, One Non-linear

Worked Example 3.12

Solve the simultaneous equations

$$2x + y = 5, \qquad 2x^2 + y^2 - xy = 7.$$

Solution

Since $2x + y = 5,$ $\qquad\qquad\qquad y = 5 - 2x.$ $\qquad\qquad\qquad\qquad$ (1)

Replace y in the second equation by $(5 - 2x)$:

$$\therefore\quad 2x^2 + (5 - 2x)^2 - x(5 - 2x) = 7,$$
$$\therefore\quad 2x^2 + 25 + 4x^2 - 20x - 5x + 2x^2 = 7,$$
$$\therefore\quad 8x^2 - 25x + 18 = 0,$$
$$\therefore\quad (x - 2)(8x - 9) = 0,$$

$\therefore\quad x = 2$; substitute in (1): $y = 1$;

or $x = \tfrac{9}{8}, y = \tfrac{11}{4}.$

Worked Example 3.13

The lift, L, produced by the wing of an aircraft varies directly as its area, A, and as the square of the airspeed, V.

    For a certain wing, $L = 1200$ when $A = 15$ and $V = 200$.

(a) Find an equation connecting L, A and V.

(b) If, for the same wing, the airspeed is increased by ten per cent, find the corresponding percentage increase in lift.   [SEB]

Solution

(a) If L varies directly as A, and as the square of V,

then $$L = kAV^2.$$

$L = 1200$ when $A = 15$, $V = 200$,

$$\therefore 1200 = k \times 15 \times (200)^2,$$

$$\therefore \quad k = \frac{1200}{15 \times 40\,000} = \frac{1}{500},$$

$$\therefore \quad L = \frac{AV^2}{500}.$$

(b) If V increases by 10 per cent, then $V = 220$.

$$\therefore L = \frac{15 \times (220)^2}{500} = 1452,$$

$$\therefore \text{Percentage increase} = \frac{\text{Increase}}{\text{Original}} \times 100$$

$$= \frac{252}{1200} \times 100 = 21\%.$$

Check 18

1 Use the remainder theorem to factorize:
    (a) $x^3 - 1$ (b) $x^3 + 2x^2 - x - 2$ (c) $x^3 + 7x^2 - 3x - 30$
    (d) $x^3 - 6x^2 + 11x - 6$ (e) $2x^3 - x^2 - 2x + 1$
    (f) $3x^3 + 8x^2 + 3x - 2$ (g) $x^3 - 6x^2 - x + 30$

2 If y is proportional to the square of x, and $y = 2$ when $x = 4$, find y when $x = 3$.

3 If p varies inversely as the square of q and $p = 4$ when $q = 2$, find p when $q = 6$.

4 If y varies as the square of x, and inversely as the cube of H, and $y = 4$ when $H = 6$ and $x = 2$, find y when $H = 4$ and $x = 6$.

5 If $x^3 + 2x^2 + ax + b$ leaves a remainder 2 when divided by $(x - 1)$ and a remainder 3 when divided by $(x + 1)$ find a and b.

Exercise 3

1 The quadratic equation with solutions $2\frac{1}{2}$ and -3 is:

    A $x^2 - 0.5x - 7.5 = 0$ **B** $x^2 + 0.5x + 7.5 = 0$

    C $2x^2 - x + 15 = 0$ **D** $2x^2 + x - 15 = 0$

2 If $t = \dfrac{x}{y} + c$, then y equals:

    A $\dfrac{x}{t + c}$ **B** $\dfrac{x}{t} - \dfrac{x}{c}$

    C $\dfrac{x}{t - c}$ **D** $\dfrac{x}{t} + \dfrac{x}{c}$

3 $\dfrac{3}{x-1} + \dfrac{4}{x+1}$ simplifies to:

   A $\dfrac{7}{x}$ **B** $\dfrac{7}{x^2-1}$ **C** $\dfrac{7x-1}{x^2+1}$ **D** $\dfrac{7x-1}{x^2-1}$

4 If $x-4$ is a factor of $x^2 - 3kx + 28$, then k equals:

   A $-3\frac{2}{3}$ **B** $3\frac{2}{3}$ **C** $9\frac{1}{3}$ **D** $-9\frac{1}{3}$

5 If $1 \leqslant a \leqslant 4$ and $2 \leqslant b \leqslant 3$, then the smallest value of $\dfrac{a}{a+b}$ is:

   A $\frac{2}{3}$ **B** $\frac{4}{7}$ **C** $\frac{1}{4}$ **D** $\frac{1}{3}$

6 If $x = 2y + 1$, then $x^2 - 2xy - y^2$ simplifies to:

   A $-y^2 - 2y + 1$ **B** $-y^2 + 2y + 1$

   C $-y^2 + 6y + 1$ **D** $-y^2 - 6y + 1$

7 If $\dfrac{1}{u} + \dfrac{1}{v} = \dfrac{1}{f}$, then if $u + v = 2$, f equals:

   A 2   **B** $\frac{1}{2}$ **C** $\dfrac{uv}{2}$ **D** none of these

8 If $\dfrac{x-2y}{x+2y} = \dfrac{4}{5}$, then $\dfrac{x}{y}$ equals:

   A 18   **B** $\dfrac{1}{18}$ **C** $4\frac{1}{2}$ **D** $\frac{1}{4}$

9 If $y = \dfrac{5x - 3}{2}$,

   then I $x = \dfrac{2y}{5} + 3$ II   when $y = 0$, $x = \frac{3}{5}$

      III $x = \dfrac{2y + 3}{5}$ IV $x = (2y + 3) - 5$

   A II only   **B** IV only   **C** II and III only   **D** II and IV only

                                                    [OLE]

10 If the solution set of the equations $x + 4y = 6$ and $3x + ky = 9$ is ϕ, then k equals:

   A 4   **B** 12   **C** $\frac{4}{3}$ **D** none of these

11 The table gives some values of u and the corresponding values of T:

u	2	4	5
T	10	2.5	1.6

   Which of the following could be the relation connecting T and u?

   A $T = \dfrac{78 - 14u}{5}$ **B** $T = \dfrac{20}{u}$ **C** $T = \dfrac{70 - 15u}{4}$ **D** $T = \dfrac{40}{u^2}$

                                                      [NISEC]

12 When a container is filled with water, the whole weighs 32 kg. When one-third of the water has been poured away, the weight becomes 24 kg. What will be the weight when one-half of the water is left?

   A 20 kg   **B** 18 kg   **C** 16 kg   **D** 12 kg   [AEB]

13 Solve the equation $\dfrac{x}{4} = \dfrac{9}{x}$.

14 If $ax = a + by^2$, express a in terms of b, x and y in its simplest form.

15 One factor of the expression $2x^3 - 7x^2 + 2x + 3$ is $(x - 3)$. Factorize the expression completely.

16 Factorize $4a^2 + 16ab + 16b^2 - 25c^2$.

17 If $v = u + at$ and $s = ut + \frac{1}{2}at^2$, find a formula for a which does not contain t, simplifying your answer.

18 Given that $y = a^2 - b^2$ and a and b are positive, find:
   (a) a when $y = 17$, $b = 2\sqrt{3}$;
   (b) b when $a = \sqrt{\frac{1}{3}}$ and $y = \frac{2}{9}$.

19 A fraction $\dfrac{a}{b}$ is such that adding 1 to the numerator gives the same result as subtracting one from the denominator. What is the value of $a - b$?

20 The operation $*$ is defined by:

$$x * y = 2x - 3y + 4xy.$$

   Find:
   (a) $2 * 3$ (b) $-1 * \frac{1}{2}$ (c) n if $2 * n = 0$

21 Using the formula $T = 100n + \dfrac{k}{v^2}$,
   (a) find the value of T if $n = 2$, $k = 1000$ and $v = 250$;
   (b) find the value of n if $T = 200$ and $k = 4v^2$.
   (c) express v in terms of T, n and k.

22 Factorize completely:
   (a) $4(x + 1)^2 - 3(x + 1)$
   (b) $(2a + 3b)^2 - (4a - b)^2$
   (c) $x^3 + x^2 - 6x$

23 Write down those solutions, if any, of the equation

$$(x - \sqrt{2})(3x - 1) = 0$$

   which are:
   (a) integers
   (b) rational
   (c) irrational
   (d) real   [Oxford]

24 Simplify:
   (a) $4p^2 - 2p^2$ (b) $(4x)^2 \div 2x$ (c) $2x^2 \times (3x)^2$

25 In the rectangle $ABCD$ in Fig. 3.1, $AB = x$ cm and $BC = 1$ cm. The line LM is drawn so that $ALMD$ is a square.

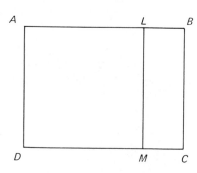

Fig. 3.1

   Write down, in terms of x,
   (a) the length LB;
   (b) the ratios $\dfrac{AB}{BC}$ and $\dfrac{BC}{LB}$.

(c) If $\dfrac{AB}{BC} = \dfrac{BC}{LB}$, obtain a quadratic equation in x. Hence find x correct to two decimal places.   [London]

26 Make x the subject of the following formulae:

(a) $y = \sqrt{\dfrac{(4x^2 + 3)t}{q}}$ (b) $hx + ky + m = 0$ (c) $4x = yt - xk.$

27 (a) Given that d varies directly as the square root of h and that $d = 7$ when $h = 4$, find
   (i) the value of d when $h = 25$,
   (ii) the value of h when $d = 14$.

(b) Show that if $3^x \times 3^{2(x^2 - 2)} = 3^4$, then the values of x are given by:

$$2x^2 + x - 8 = 0.$$

Hence, calculate the positive value of x, correct to two decimal places.

[AEB]

28 It is given that $P = 2nx - y.$
   (a) Calculate P when $n = 3$, $x = 4$ and $y = -5$.
   (b) Express y in terms of n, x and P.
   (c) Express n in terms of P, x and y.
   (d) If P, x and y are all doubled, state the effect on n.   [AEB]

In questions 29–31 the equations are all based on examination questions.

29 Find x:
   (a) $4(3 - 2x) = 5$;   (b) $2x^2 + x - 6 = 0$;
   (c) $4x(3 + x) = 0$,   (d) $2x + \dfrac{x}{3} = 5$;
   (e) $\dfrac{2 + x}{4} - \dfrac{2 - x}{4} = \tfrac{1}{2}$;   (f) $4x^2 - 100 = 0$;
   (g) $\dfrac{3}{x + 1} = 4$;   (h) $3(x + 2) - (x - 2) = 6$;
   (i) $(x - 2)(x + 2) = 5$;   (j) $\dfrac{4}{x^2} = \dfrac{x}{2}.$

30 Make x the subject:
   (a) $q - x = 4y$;   (b) $T = \sqrt{\dfrac{4}{x + 2}}$;
   (c) $y = \dfrac{4x + t}{x}$;   (d) $x^2 + y^2 = r^2$;
   (e) $x - y = \dfrac{4}{x + y}$;   (f) $A = \dfrac{4\sqrt{x^2 + y^2}}{t}$;
   (g) $ax + by + c = 0$;   (h) $E = \tfrac{1}{2}mx^2$;
   (i) $F = \dfrac{1}{x} + \dfrac{1}{f}.$

31 Factorize:
   (a) $2tq^2 - 18t$ (b) $4tq + 8t^2$
   (b) $2tx + 2ty - 3x - 3y$ (d) $32x^2 - 28x - 99$
   (e) $x^3 + x^2 - 2$

32 Evaluate the unknown quantity in each equation, using the values $a = 2$, $b = -1$, $c = -4$:
   (a) $A = 3a - 2b + c$;   (b) $B = 4a + 5b^2$;
   (c) $C = a - b^2 - 2c$;   (d) $D = 5ab - 3bc + 2ac.$

33 The volume of a closed rectangular box is 100 cm^3. The width of the box is x cm, and the height of the box is twice the width. Express in terms of x:

(a)   the length of the box,   (b)   the surface area of the box.
If the surface area equals 160 cm^2, show that a possible value of x is 5.

34 (a)   Solve the equations:
$$3x + 4y = 2,$$
$$5x - 3y = 13.$$

(b)   Given that $\dfrac{1}{u} + \dfrac{1}{v} = \dfrac{1}{f}$, express u in terms of v and f.

(c)   Express $x^2 + 8x + 25$ in the form $(x + k)^2 + t$, hence show that whatever the value of x, $x^2 + 8x + 25$ cannot equal 8.

35 Jane earned £y a week for the first w weeks of the year, and £r a week for the rest of the year. Write down an expression for her average weekly wage.

36 The pressure P at a point in a liquid is proportional to the depth h of the point below the surface. If $P = 460$ when $h = 5$, write down a relationship connecting P and h. Hence find the value of h when $P = 600$.

37 The force F between two objects is inversely proportional to the square of the distance d between them. If $F = 400$ when $d = 2$, write down a relationship connecting F and d. Use this to find
(a) the value of F when $d = 1$;
(b) the value of d when $F = 25$.

38 Kepler's third law states that the square of the time T (days) of a planet to complete one orbit of the sun is proportional to the cube of the average distance D (km) from the sun.

For the planet Venus, $T = 88$ days and $D = 58 \times 10^6$ km. Work out a formula connecting T and D. Assuming that the average distance of the earth from the sun is 1.5×10^8 km, find how many days the formula predicts that the earth takes to travel round the sun.

39 Two congruent rectangles $ABCD$ and $PQRS$ each of length l metres and breadth b metres are placed to form the capital letter 'T' shown in Fig. 3.2.

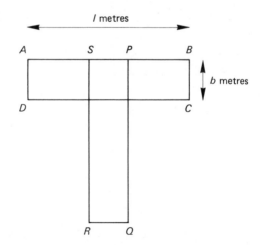

Fig. 3.2

(a)   (i) Show, with an explanation using words or a diagram, that the area A square metres of the T-shape is given by the formula $A = b(2l - b)$.
(ii) Rearrange this formula to express l in terms of A and b.

(b) Find the formula for the perimeter p metres of the T-shape.   [SEB]

40 A solid cuboid of length p metres has a square cross-section of side $(p - 3)$ metres. Write down, in terms of p, expressions for:
(a) the surface area of the square cross-section;

(b) the surface area of one of the longer faces of the cuboid.

Given that the total surface area of the cuboid is 39 square metres, write down an equation in p and show that it reduces to

$$2p^2 - 8p - 7 = 0.$$

Find the two values of p which satisfy this equation, giving your answers correct to three significant figures.

Hence write down the dimensions of the cuboid. [UCLES]

41 An aeroplane travelled a distance of 500 km at an average speed of x km/h. Write down an expression for the number of hours taken.

On the return journey the average speed was increased by 10 km/h. Write down an expression for the number of hours taken on the return journey.

Given that the time taken for the return journey was 5 *minutes* less than that of the outward journey, form an equation in x and show that it reduces to

$$x^2 + 10x - 60\,000 = 0.$$

Given that $x^2 + 10x - 60\,000$ can be expressed in the form $(x + 250)(x - p)$, find the value of p.

Hence find the time taken on the outward journey. [UCLES]

4 The Metric System

4.1 Units

The standard unit of length is the *metre* (m).
   1 kilometre (km)   = 1000 m
   1 millimetre (mm) = 0.001 m
   1 centimetre (cm) = 0.01 m
                1 cm = 10 mm
   1 square metre will be written 1 m²
   1 cubic metre will be written 1 m³
   1 hectare = 10 000 m²
   1 litre = 1000 cm³
The standard unit of mass is the kilogram (kg).
   1 kg   = 1000 grams
   1 gram (g)   = 0.001 kg
   1 milligram (mg) = 0.001 g
The standard unit of time is the second (s).
Speeds, for example 8 metres per second, are written 8 m/s in this book (rather than 8 m s⁻¹).

4.2 Changing Units in Area and Volume

How many mm² are there in 1 cm²?
A very common but incorrect answer is 10.
   Figure 4.1 shows an enlarged centimetre square divided into square millimetres (shaded). It can be seen that the answer is $100 = 10^2$.

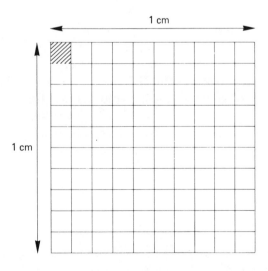

Fig. 4.1

When changing units of area, the multiplication factor is the square of the multiplication factor for the unit of length.

Similarly, when changing units of volume, we find that the multiplication factor is the cube of the multiplication factor for the unit of length.

4.3 Conversion of Units Involving Rates

How can we change 80 m/s into km/h?

Write 80 m/s as $\dfrac{80\,\text{m}}{1\,\text{s}}$.

But 1000 m = 1 km, $\therefore$ 1 m = $\frac{1}{1000}$ km.

3600 s = 1 h, $\therefore$ 1 s = $\frac{1}{3600}$ h.

Hence $\dfrac{80\,\text{m}}{1\,\text{s}} = \dfrac{80 \times \frac{1}{1000}\,\text{km}}{1 \times \frac{1}{3600}\,\text{h}} = \dfrac{80}{1000} \times \dfrac{3600\,\text{km/h}}{1}$

$$= 288\,\text{km/h}.$$

Another example: how can we change 80 g/cm³ into kg/m³ ?

Write 80 g/cm³ as $\dfrac{80\,\text{g}}{1\,\text{cm}^3}$.

1000 g = 1 kg, $\therefore$ 1 g = $\frac{1}{1000}$ kg.

1 m³ = (100)³ cm³, $\therefore$ 1 cm³ = $\frac{1}{100\,000}$ m³,

$$\therefore \dfrac{80\,\text{g}}{1\,\text{cm}^3} = \dfrac{80 \times \frac{1}{1000}}{1 \times \frac{1}{1\,000\,000}} = 80\,000\,\text{kg/m}^3 .$$

Check 19

1 Express the following lengths in cm.
   (a) 4.3 m   (b) 684 mm   (c) 400 m   (d) 0.05 m
   (e) 2.5 mm   (b) 863 km   (g) 0.04 km (h) 1255 mm
2 Express the following lengths in km:
   (a) 625 cm   (b) 8000 cm   (c) 9400 mm (d) 895 m   (e) 86.9 cm
3 Express the following masses in g:
   (a) 8 kg   (b) 0.04 kg   (c) 685 mg   (d) 65 kg   (e) 1.8 mg   (f) 900 kg
4 Express the following areas in cm² :
   (a) 9.6 m²   (b) 2875 mm²   (c) 0.048 km²
5 Change the following volumes into litres:
   (a) 600 cm³   (b) 28 m³   (c) 2 km³
6 Carry out the following conversions:
   (a) 20 m/s into km/h   (b) 25 km/h into m/s   (c) 8 g/cm³ into kg/l

4.4 Area

A list of formulae for the areas of a number of useful shapes is given in Fig. 4.2. Proofs are not given as these can be found elsewhere.

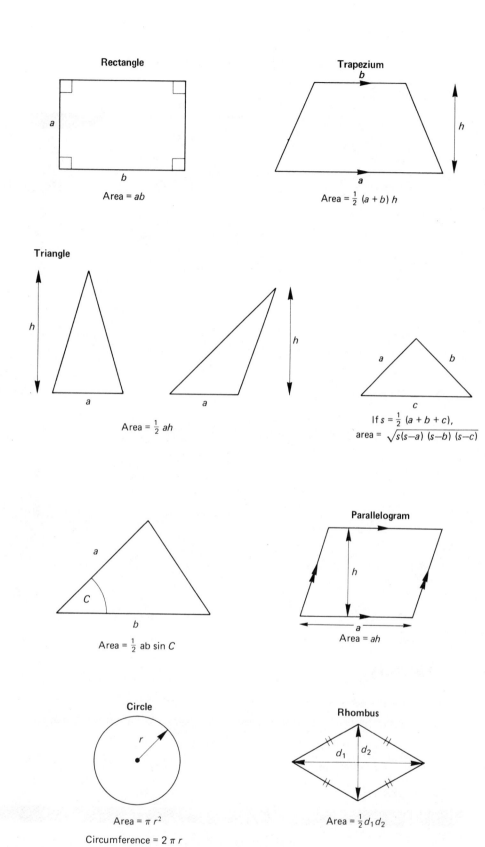

Fig. 4.2

4.5　Volume

Formulae for finding volumes are given in Fig. 4.3.

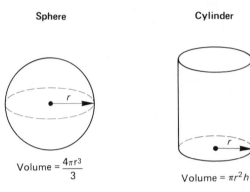

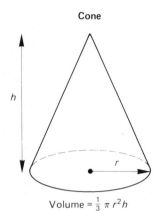

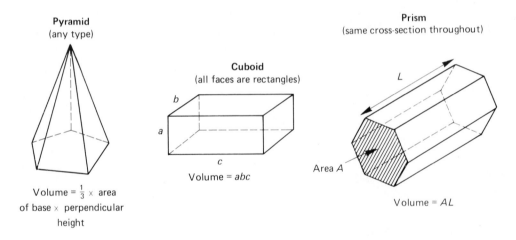

Fig. 4.3

4.6　Density

The mass of a unit volume of a given substance is called its *density*.

$$\text{Density} = \frac{\text{Mass}}{\text{Volume}}.$$

Units are kg/m³ etc.

See Worked Example 4.2.

4.7　Surface Area

Referring again to Fig. 4.3,

Sphere: surface area = $4\pi r^2$.

Cylinder: curved surface area = $2\pi rh$.
total surface area including the ends
$$= 2\pi r^2 + 2\pi rh = 2\pi r(r+h).$$

Cuboid: surface area = $2ab + 2bc + 2ac$.

Cone: we need to assume that the vertex of the cone is above the centre of the base. If the slanting edge of the cone is of length l (Fig. 4.4), then the curved surface area of the cone = πrl.

Total surface area = $\pi r^2 + \pi rl$
$$= \pi r(r+l).$$

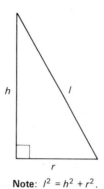

Note: $l^2 = h^2 + r^2$.

Fig. 4.4

When using any of the formulae for area and volume, it is important (in fact, necessary), that all measurements are in the same units.

Practice using the following check exercise.

Check 20

1 Find the area of the following shapes; all measurements are in cm ($\pi = 3.14$).

(a)

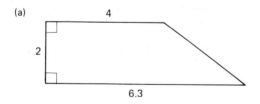

(b)

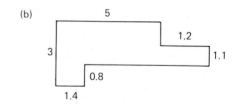

(c)

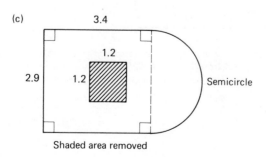

Shaded area removed

(d)

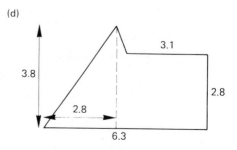

Fig. 4.5

69

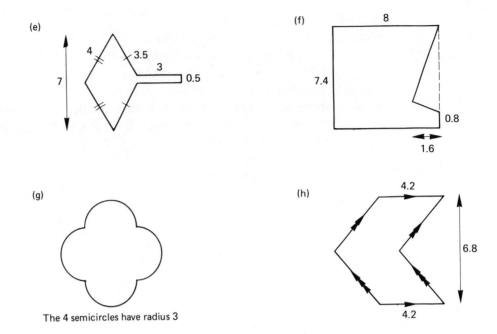

(e)

(f)

(g) The 4 semicircles have radius 3

(h)

Fig. 4.5 cont.

2 A cuboid shaped box measuring 2 m × 85 cm × 1.6 m is packed with small $\frac{1}{2}$ cm cubes. How many can be packed into the box?

3 Find the area of a regular hexagon with sides of length 6 cm.

4 The sides of a rectangle are of length $3x$ and $4x$ cm. Given that the area of the rectangle is 147 cm², find x.

5 The following questions involve the use of $\pi = \frac{22}{7}$ in the calculations. All answers should be expressed as fractions. Copy and complete the tables:

Cylinder:

Volume	Height	Radius	Total surface area
	4 cm	7 cm	
176 m³		2 m	
	70 cm	1 m	
		14 cm	1408 cm²

Sphere:

Volume	Surface	Radius
		21 cm
$33\frac{11}{21}$ m³		
	9856 cm²	
		1.4 cm
	98.56 m²	

Cone:

Volume	Curved surface area	Total surface area	Base radius	Height	Slant height
			7 cm	4 cm	
				12 cm	13 cm
154 cm³				3 cm	
	704 m²		8 m		
		1496 cm²	14 cm		

70

Worked Example 4.1

Figure 4.6 shows a horizontal rectangle *ABCD* where *AB* = 2 m and *AD* = 1 m, which forms the open top of a water container of semicircular cross-section. Calculate the capacity of the water container (a) in cubic metres; (b) in litres, giving your answer in each case to 3 significant figures.

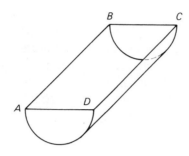

Fig. 4.6

The water container, including the ends, is to be made out of sheet metal. Calculate the area of sheet metal required in square metres, correct to three significant figures.

Calculate also the area of the surface of the water when the depth of the water at the deepest point is 0.2 m, giving your answer in square metres (π = 3.14).

[OLE]

Solution

(a)   The shape is best regarded as half a cylinder.

$$\therefore \quad \text{Volume} = \tfrac{1}{2}\pi r^2 h = \tfrac{1}{2} \times 3.14 \times 0.5^2 \times 2$$

Radius is $\tfrac{1}{2}AD$ *AB* is height

$$= 0.785 \text{ m}^3$$

(b)   1 litre = 1000 cm³.

1 m³ = (100)³ cm³ = 1 000 000 cm³ = 1000 litres,

$\therefore$ capacity is 785 litres.

The surface area of the container is also half the total surface area of a cylinder.

$$\therefore \quad \text{Area of metal} = \tfrac{1}{2} \times 2\pi r(r + h) = \pi r(r + h)$$
$$= 3.14 \times 0.5 \times 2.5 = 3.93 \text{ m}^2.$$

To find the surface area of the water when the depth is 0.2 m, we need to find the distance *XY* shown in Fig. 4.7.

Since *PQ* = 0.3 m and *XP* = 0.5 m (radius),

XQ = 0.4 ($\triangle$ *XPQ* is a 3, 4, 5 triangle).

$\therefore$ Surface area = 0.8 × 2

$$= 1.6 \text{ m}^2.$$

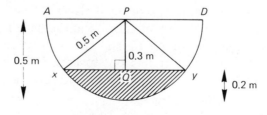

Fig. 4.7

Worked Example 4.2

A roll of copper wire is made of wire 200 m long with a circular cross-section of diameter 1.2 mm. Calculate the volume of the coil.

If the density of copper is 8.8 g/cm³, calculate the weight of the copper wire in kilograms.

Solution

Care must be taken with the units. Since the density is given per cubic centimetre, it is best to calculate the volume in cubic centimetres.

Using $V = \pi r^2 h$,

$$V = 3.14 \times 0.06^2 \times 20\,000$$
$$= 226 \text{ cm}^3.$$

$$\therefore \text{ Weight } = \frac{226 \times 8.8}{1000} \text{ kg}$$

$$= 1.989 \text{ kg}.$$

Worked Example 4.3

Figure 4.8 shows the vertical front face of a greenhouse. (The back face of the greenhouse is identical to the front face and the base of the greenhouse is rectangular.) $\triangle ABE$ is isosceles, with $AB = AE = 1.0$ m, and $BCDE$ is a trapezium with $BC = ED = 1.4$ m, $BE = 1.5$ m and $CD = 1.9$ m.

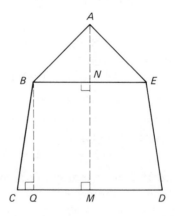

Fig. 4.8

Calculate, to two decimal places,
(a) the height AN of $\triangle ABE$ and the distance NM;
(b) the area of $\triangle ABE$ and the area of the trapezium $BCDE$. The length of the greenhouse from front to back is 2.5 m and the whole of each of its six faces (front, back, sloping sides and roof) is covered with glass.

Find:
(c) the total area, in m², of glass;
(d) the volume, in m³, of air inside the greenhouse.   [L]

Solution

(a) By Pythagoras' theorem,

$$AB^2 = AN^2 + NB^2,$$

$$\therefore \quad 1^2 = AN^2 + (0.75)^2,$$

$$\therefore \quad AN = \sqrt{1 - 0.5625} = 0.66 \text{ m}.$$

BQ is a construction line, not on the original diagram.

$$\therefore \quad CB^2 = BQ^2 + CQ^2 = NM^2 + (0.2)^2,$$

$$\therefore \quad 1.4^2 - 0.2^2 = NM,$$

$$\text{hence} \quad NM = 1.39 \text{ m}.$$

(b) Area of $\triangle ABE = \frac{1}{2} \times 1.5 \times 0.66 = 0.5 \text{ m}^2$.
Area of trapezium $BCDE = \frac{1}{2}(1.9 + 1.5) \times 1.39 = 2.36 \text{ m}^2$.
(c) The surface consists of two rectangles $1 \times 2.5 = 5 \text{ m}^2$
two rectangles $1.4 \times 2.5 = 7 \text{ m}^2$.
Two ends $2.36 + 0.5 = 5.72 \text{ m}^2$.
$\therefore$ Total area of glass $= 17.72 \text{ m}^2$.
To calculate the volume, the greenhouse can be regarded as a prism.

$$\text{Volume} = \text{Area of end} \times 2.5$$
$$= 2.86 \times 2.5$$
$$= 7.15 \text{ m}^3.$$

4.8   Approximating Areas, Trapezium Rule

There are many occasions when an area (or volume) cannot be found exactly by formula.

The following technique, known as the *trapezium rule*, can be used in many situations (see section 4.13 on velocity- (speed-) time graphs).
In order to find the shaded area in Fig. 4.9 it has been divided into four trapezia (any number can be used). The width of each trapezium is h. The y values y_1, $y_2, \ldots$, give the length of the sides.

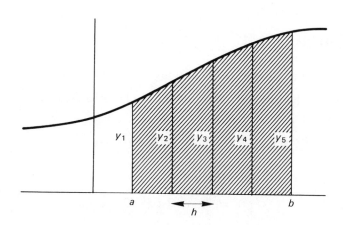

Fig. 4.9

$$\text{Total area} = \frac{h}{2}(y_1 + y_2) + \frac{h}{2}(y_2 + y_3) + \frac{h}{2}(y_3 + y_4) + \frac{h}{2}(y_4 + y_5).$$

Factorizing: $= \dfrac{h}{2}(y_1 + 2y_2 + 2y_3 + 2y_4 + y_5).$

This formula is easily extended to any number.
See Worked Example 4.5.

4.9   Radian Measure

In more advanced work, the idea of measuring angles in degrees is often not very useful; a much better way is to use what is called *radian* measure.

Figure 4.10 shows a sector of a circle, with arc length equal to the radius. The angle at the centre is defined as 1 radian, sometimes written 1^c or 1 rad.

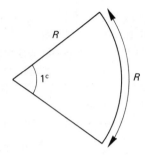

Fig. 4.10

Since the circumference of the circle is $2\pi R$ then the fraction of a complete circle is

$$\frac{R}{2\pi R} = \frac{1}{2\pi},$$

hence there are 2π radians in $360°$, or

$$\boxed{\pi^c = 180°.}$$

Worked Example 4.4

(a) Change the following angles into radians: $60°, 45°, 112°$.

(b) Change the following angles into degrees: $\dfrac{\pi^c}{6}, \dfrac{\pi^c}{2}, 1.3^c$.

Solution

(a) Since $180° = \pi^c$,

$\div 3$: $60° = \dfrac{\pi^c}{3}$ or 1.05^c (2 d.p.);

$\times \frac{3}{4}$: $45° = \dfrac{\pi^c}{4}$ or 0.79^c (2 d.p.).

Using the ideas developed in section 2.11,

$$180° = \pi^c.$$

$$\div 180: 1° = \frac{\pi^c}{180}.$$

$$\times 112: 112° = \frac{\pi}{180} \times 112^c = 1.95^c.$$

(b) Since $\pi^c = 180°$,

$$\div 6: \frac{\pi}{6}^c = 30°,$$

$$\text{or} \div 2: \frac{\pi}{2}^c = 90°.$$

Once again, using section 2.11:

$$\pi^c = 180°.$$

$$\div \pi: 1^c = \frac{180°}{\pi}.$$

$$\times 1.3: 1.3^c = \frac{180}{\pi} \times 1.3 = 74.5° \text{ (1 d.p.)}.$$

4.10   Sectors and Segments

Referring to Fig. 4.11,

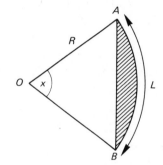

Fig. 4.11

(a) Angle x in degrees:

length of arc $L = 2\pi R \times \dfrac{x}{360}$,

area of sector $OAB = \pi R^2 \times \dfrac{x}{360}$,

area of segment (shaded) $= \dfrac{\pi R^2 x}{360} - \frac{1}{2}R^2 \sin x°$.

This is found from area of sector *minus* area of triangle.

(b) Angle x in radians.

length of arc $L = 2\pi R \times \dfrac{x}{2\pi}$

(since $360°$ is $2\pi^c$);
this simplifies to $L = Rx$.

Area of sector $OAB = \pi R^2 \times \dfrac{x}{2\pi}$;

this simplifies to $\frac{1}{2}R^2 x$.
Area of segment $= \frac{1}{2}R^2 x - \frac{1}{2}R^2 \sin x$
$= \frac{1}{2}R^2 (x - \sin x)$.

These formulae are much simpler, but care must be taken with the segment formula, that the calculator is switched to radians before using the sine button.

Check 21

1 Use the trapezium rule with 5 strips to find the area under the curve $y = x^2 + 1$ between $x = 1$ and $x = 2$.

2 The values in the table give the relationship between x and t:

t	0	1	2	3	4	5
x	0.8	1.3	1.8	2	1.4	0.8

Use the trapezium rule to find the area under the graph of x plotted against t.

3 Change into radians:
   (a) $45°$; (b) $60°$; (c) $180°$; (d) $135°$; (e) $24°$; (f) $7\frac{1}{2}°$.

4 Change the following into degrees:
   (a) 0.3^c; (b) 3^c; (c) 0.86^c; (d) $\frac{\pi}{8}^c$; (e) $\frac{2\pi}{3}^c$; (f) 3.14^c.

5 Calculate the lengths of arcs of the following:

Radius	Angle subtended at centre
(a) 7 cm	$70°$
(b) 6370 km	$130°$
(c) 5.7 m	$4° \, 30'$
(d) 6400 km	$22° \, 30'$

6 Calculate the radii of the circles for the following:

Arc length	Angle subtended at centre
(a) 5 cm	$20°$
(b) 7.3 cm	$13°$
(c) 825 km	$75°$
(d) 400 km	$112° \, 30'$

7 Calculate the angle subtended by the following arcs to the nearest degree:

Radius	Length of arc
(a) 8 cm	12 cm
(b) 6370 km	6370 km
(c) 528 m	49 m
(d) 6370 km	150 km

4.11   Speed, Velocity and Acceleration

Speed is defined as $\dfrac{\text{distance travelled}}{\text{time taken}}$.

More specifically, *average* speed = $\dfrac{\text{total distance travelled}}{\text{total time taken}}$.

Units are m/s, km/h, etc.

Velocity is a *vector*. It describes both the speed and the direction of travel given.

$Acceleration = \dfrac{\text{change in velocity}}{\text{time taken}}$. Slowing down is called *retardation* (negative acceleration).

4.12   Displacement − Time (or Distance − Time) Graphs

If we plot a graph of displacement (distance from a point) against time, as in Fig. 4.12, it can be read as follows:

OA: Straight line means constant speed.
AB: Horizontal line means at rest.
BC: Gradually increasing and then decreasing gradient, means acceleration followed by retardation.
CD: At rest.
DEF: Constant speed returning to *O* and then going beyond *O* for *EF*.
FG: Return to *O* at constant speed.

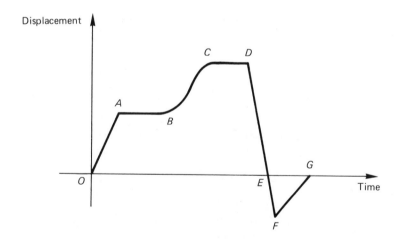

Fig. 4.12

4.13   Velocity (Speed) − Time Graphs

A velocity–time graph (see Fig. 4.13) is interpreted as follows:

OA: Straight line, constant acceleration.
AB: Horizontal line, constant speed (acceleration zero).
BC: Velocity decreasing, not straight, therefore non-uniform retardation.

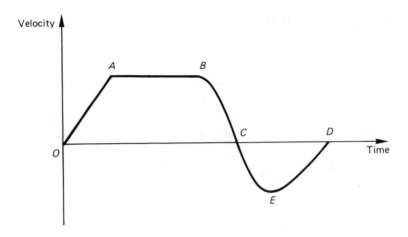

Velocity

A   *B*

O   *C*   *D*   Time

E

Fig. 4.13

CED: Stops at *C*, increases speed still on opposite direction until *E* is reached, then decreases until it stops at *D*. Non-uniform acceleration.

The area under a velocity–time graph gives the distance travelled. If the graph does not consist of straight lines, the trapezium rule can be used.

Worked Example 4.5

The table below gives the velocity v (m/s) of a body measured every 2 s over a 10 s period. Estimate the distance travelled in that time.

t	0	2	4	6	8	10
v	0	3	7	15	24	20

Solution

The graph would not be a straight line.
∴ Using the trapezium rule, with $h = 2$,

$$\text{Distance} = \frac{2}{2} \ (0 + 2 \times 3 + 2 \times 7 + 2 \times 15 + 2 \times 24 + 20)$$

$$= 118 \text{ m}.$$

Worked Example 4.6

The diagram, Fig. 4.14, shows the speed–time graph for an underground train travelling between two stations. It starts from rest and accelerates at a constant rate for 10 seconds, then travels at constant speed for 30 seconds, and finally slows down at a constant rate. If the distance between the stations is 850 m, calculate:
(a) the maximum speed; (b) the acceleration; (c) the time it takes to reach the half-way point.

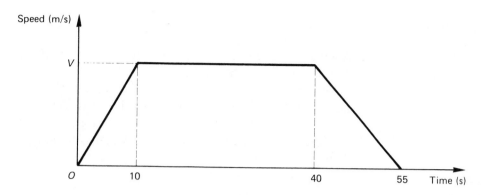

Fig. 4.14

Solution

Many questions on distance–time graphs can be solved on the basis that the area under the graph equals the distance travelled. Look out for triangles, trapeziums, etc.

(a) Area of the trapezium $= \frac{1}{2}(55 + 30)v$

$$= \frac{85v}{2}.$$

But distance = area, $\therefore \frac{85v}{2} = 850$;

this gives $v = 20$.

$\therefore$ Maximum speed = 20 m/s.

(b) The acceleration equals the gradient of the line.

$\therefore$ Acceleration $= \frac{20}{10} = 2$ m/s^2.

(c) To find how long it takes to reach the half-way point, we are trying to find the time T which divides the area under the graph into two equal parts; see Fig. 4.15. The shaded area must equal $\frac{850}{2} = 425$.

Since the triangle Ⓐ has area $= \frac{1}{2} \times 10 \times 20 = 100$, then rectangle Ⓑ has area $425 - 100 = 325$;

$\therefore (T - 10) \times 20 = 325$.

$T - 10 = \frac{325}{20} = 16.25$; $\therefore T = 26.25$.

$\therefore$ It takes 26.25 seconds.

Note: This is not one-half of 55 seconds.

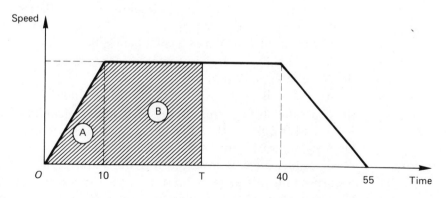

Fig. 4.15

Exercise 4

1 A piece of wire 4 m long has a volume of 0.4 cm³. Its cross-sectional area is:
   A 0.01 cm²   B 0.1 mm²   C 0.4 mm²   D none of these
2 A bucket is in the form of a cylinder of radius 20 cm, and height 40 cm. If the mass of 1 litre of water is 1 kg, then the mass of water in the bucket is:
   A 16π kg   B 160π kg   C 1.6π kg   D 16 kg
3 The diagram (Fig. 4.16) shows the speed–time graph of an object. The total distance travelled is:

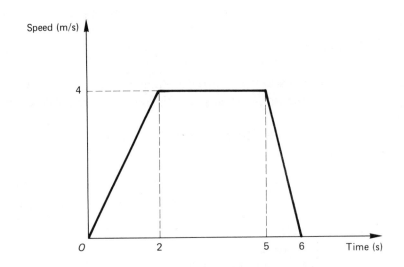

Fig. 4.16

   A 20 m   B 24 m   C 16 m   D 18 m
4 A cube has side of length 10 cm. A and B are two vertices of the cube which are furthest apart. A fly walks along the surface of the cube from A to B. The shortest distance it can walk is:
   A $10\sqrt{5}$ cm   B $10\sqrt{3}$ cm   C 30 cm   D $20\sqrt{2}$ cm
5 A sector of a circle has radius 10 cm, and length of arc 20 cm. The angle at the centre of the sector measured in radians is:
   A 2π B 2   C 4   D π
6 A car increases its speed from 20 m/s to 40 m/s while travelling a distance of 100 m. Its acceleration is:
   A 0.2 m/s²   B 6 m/s²   C 6 m/s   D none of these
7 The radius of a circle whose circumference is 50π cm is
   A 25 cm   B $12\frac{1}{2}$ cm   C 25π cm   D $12\frac{1}{2}\pi$ cm
8 The total surface area of a solid hemisphere of radius 7 cm is:
   A 245 cm²   B 770 cm²   C 245π cm²   D none of these
9 72 km/h expressed in m/s is:
   A 2   B 20   C 200   D 2000
10 The total surface area of a solid cylinder of radius 6 cm and height 2 cm is:
   A 144π cm   B 108π cm²   C 144π cm²   D none of these
11 A cylinder of radius 6 cm contains water to a depth of 4 cm. A small sphere of radius 2 cm is lowered into the water slowly to avoid spillage. Without using a numerical value for π, find the exact change in the level of water.
12 The length of the semicircular arc BEC in Fig. 4.17 is 4π cm. Calculate the area of the square $ABCD$.   [OLE]

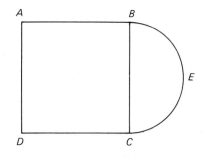

Fig. 4.17

13 A rectangular cuboid measures 8 cm by 40 cm by 85 cm. Find, giving your answers in standard form:
(a) its surface area in mm² ;
(b) its volume in m³ .

14 In Fig. 4.18, the radius of the larger circle is twice that of the smaller concentric circle and θ is the angle between two radii of the larger circle.

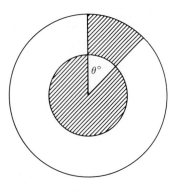

Fig. 4.18

(a) Find the value of θ for which the two shaded regions have equal areas.
(b) Find the value of θ for which the two shaded regions have equal perimeters.

[NISEC]

15 A county council surveyor calculating the costs of road construction determines the costs from the following:
(a) preparation, etc. £9 000 per km;
(b) excavation: £300 000 per km;
(c) materials: hardcore £310 per m³ ;
base sealer £150 per m³ ;
final surface £170 per m³ .
Construction regulations require the hardcore to be 20 cm thick, the base sealer to be 8 cm thick and the final surface to be 6 cm thick.

Find the total costs involved in constructing each km of a new road 8 m wide.

For resurfacing roads of the same width, no other costs apart from the final surfacing costs are incurred. If, in a year, the County Council decides to spend £42 432 000 on this task, calculate the total length of roads of this width which may be resurfaced. [L]

16 In a certain year a factory producing coats had a total wage bill of £1 500 000. Of this bill, 20 per cent was paid for overtime and the remainder for normal working. The total amounts paid for normal working to each of three groups

of employees, namely management, skilled workers and semi-skilled workers, were in the ratios 1 : 6 : 3. Calculate the yearly wages bill for normal working for each of the three groups of workers.

The average payments for normal work to employees in each of the three groups were in the ratio 5 : 3 : 2, and the factory employed 20 persons in the management group. Using the wages bill as calculated above, calculate:

(a) the average yearly wage, for normal working, for an employee in each of the three groups;

(b) the total number of persons employed in the factory.

The payments for overtime were distributed between the three groups of employees in the ratio 0 : 3 : 2. Calculate the *total* average yearly wage of a person in the skilled worker group and express this as a percentage of the *total* average yearly wage of a person in the management group. [L]

17 The diagram in Fig. 4.19, not drawn to scale, shows the uniform vertical cross-section *PQRST* of a swimming pool of length 50 m and width 12 m. The deepest point *S* of the cross-section is vertically below the point *V*. The areas of the cross-section *PVST* and *VQRS* are equal and *PT* = 1 m, *VS* = 6 m and *QR* = 2 m. Given that *VQ* = x m, show that $x = 23\frac{1}{3}$.

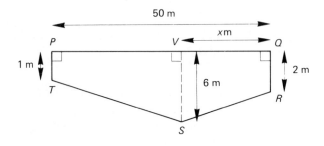

Fig. 4.19

Calculate the volume of water in the pool:

(a) when it is completely full;

(b) when the whole of the surface of the base of the pool is just covered with water.

(c) The pool is empty and is then filled by water flowing at a rate of $2\frac{3}{4}$ m/s through a system of 10 cylindrical pipes. Given that the radius of the cross-section of each of the pipes is r cm, and taking π as $\frac{22}{7}$, write down an expression in terms of r for the volume, in m³, of water flowing into the pool each hour through the system of pipes. Given that it takes 16 hours to fill the pool completely, calculate the value of r. [L]

18 To estimate the cost of providing windows, glazing suppliers measure the perimeter of the window spaces in millimetres and charge the following rates per mm:

1.25p per mm for glazing with a single piece of glass;

1.5p per mm for glazing with two sliding pieces of glass;

1.75p per mm for glazing with three sliding pieces of glass.

I have four windows which each measure 1200 mm by 1780 mm.

(a) Calculate the perimeter of one window.

I decide to fit two sliding pieces of glass to each of two of the windows.

(b) Calculate the cost of glazing these two windows.

I decide to fit three sliding pieces of glass to each of the other two windows.

(c) Calculate the cost of glazing these two windows.

In addition, I decide that a single pane of glass will be adequate for the small hall window which measures 1050 mm by 840 mm.

(d) Calculate the cost of glazing this window.

(e) Find the total cost of all five windows.

An extra 10 per cent is added to this total for fitting the windows.

(f) Calculate the fitting charge.

A discount of 5 per cent of the total price (including the fitting charge) is allowed for a cash payment.

(g) Find, to the nearest penny, what saving this would represent on this order.

[L]

19 A body starts from rest, and moves with constant acceleration for 3 s. It then moves with constant speed x m/s for 6 s, and finally comes to rest with constant retardation. The complete journey takes 13 s.

(a) Draw the velocity–time graph for the journey.

(b) Find the total distance travelled in terms of x.

(c) Find x, if the average speed for the journey is 38 m/s.

20 Calculate the amount of liquid that can be stored in a rectangular tank measuring 4 m by 8 m by 1.5 m.

The tank is filled using a pipe of cross-sectional area 0.08 m^2, at a constant speed of 2 m/s. Calculate the time it takes to fill the tank.

The tank is then emptied into a similar tank, the smallest dimension being 3 m. Find the volume of this second tank, and the depth of the liquid, assuming that 3 m is the height of the second tank.

21 The speed v m/s of an experimental trolley is measured running along a straight track after t seconds, and found to satisfy the equation $v = 16t - t^2$.

(a) Draw up a table of values for $0 \leqslant t \leqslant 16$ at intervals of 2 seconds.

(b) Using the scales of 1 cm to represent 5 m/s vertically and 1 cm to represent 2 seconds horizontally, draw the graph for $0 \leqslant t \leqslant 16$.

(c) Using your graph, (i) find the maximum speed of the trolley; (ii) find the acceleration of the trolley when $t = 4$ seconds; (iii) estimate the distance travelled in the 16 seconds, using the trapezium rule. Show your working clearly.

22 The diagram in Fig. 4.20 shows the speed–time graph of a car starting from rest, and travelling for 16 seconds with constant acceleration. Its speed at the end of this time is v m/s. If the total distance travelled is 480 m, calculate:

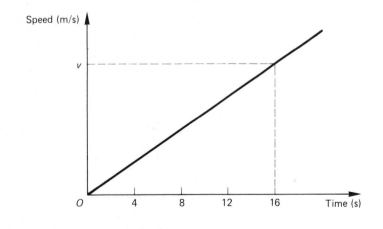

Fig. 4.20

(a) the value of v;

(b) the time it took to cover the first 240 m;

(c) the acceleration of the car.

23 A train travels between two towns A and B, 60 km apart. The scheduled time for the journey is 1 hour. The first two-thirds of the journey is travelled non-stop at an average speed of 80 km/h. There are three stations in the last one-third of the journey, and the average stop at each station is x min. If the average speed for the last part of the journey is 60 km/h, find x. If the maximum amount that x can be reduced by is 1 min and the train is 10 min late after the first two-thirds of the journey, what is the earliest time the train can arrive at B, assuming that the average speed for the last part of the journey is the same as before?

24 The diagram (Fig. 4.21) shows a glass funnel filled with water. Calculate the volume of water it can hold. Calculate the percentage error that is made if you assume the conical part comes to a point where it joins the cylindrical part.

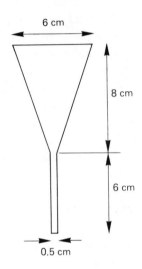

6 cm

8 cm

6 cm

0.5 cm

Fig. 4.21

If half the water is allowed to run out of the bottom of the tube, calculate the drop in the level of water, assuming the top does come to a point.

25 A family man wishes to hire a car for a day. Three car-hire firms quoted their prices as follows:

Alpha Agency charged £20 per day irrespective of the distance travelled.
Betta Travel charged £10 for any distance up to, and including 200 km, and then 6p for every kilometre over the initial 200 km.
Cruiser Car charged a basic £5 plus 4p for every kilometre travelled.

(a) Draw up a table that shows the cost of hiring a car for one day from each firm and using it to travel 0, 200 and 400 km in that day.

(b) Use your table to draw graphs on the same axes showing the cost of hiring a car from each firm for all distances up to 400 km in one day.

(Scales: take 2 cm to represent 50 km on the distance axis and 2 cm to represent £2 on the cost axis.)

Use your graphs to estimate:

(c) the range of distances for which Betta Travel is cheaper than Cruiser Cars;

(d) the range of distances for which Alpha Agency is cheaper than either of the other two firms;

(e) the maximum cost per kilometre that Cruiser Cars would have charged if the firm kept its basic £5 charge and was cheaper than both the other two firms up to distances of 400 km. [AEB]

5 Functions

5.1   Definition

Figure 5.1 shows the relationship between the elements of a set Y, and those of a set X. It can be seen that if $y \in Y$ and x is the value it is related to in X (shown by an arrow), then $y = x + 2$. This mathematical 'rule' which changes a value of x into a value of y is called a *function* or *mapping*. It can also be written $f : x \longmapsto x + 2$.

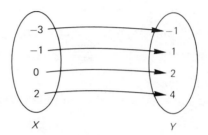

Fig. 5.1

The function f maps x onto $x + 2$. The letter f is not important, and a different letter should be used for a different function. To find the effect of f on a particular value of x, say $x = 3$, would be written f(3) (read as 'function of 3'). It follows that: $f(3) = 3 + 2 = 5$.

We can also write f as $f(x) = x + 2$. Since f describes a rule, the variable x is not important, and is called a *dummy* variable. f could also be written as $f : t \longmapsto t + 2$ or $f(y) = y + 2$.

5.2   Domain, Codomain, Range, Image

The above terms are illustrated by Fig. 5.2. The set of values used for x is called

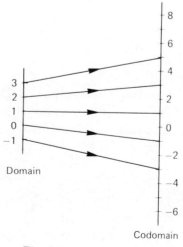

Fig. 5.2

85

the *domain* of the function. In this case, the domain = {−1, 0, 1, 2, 3}. The *co-domain* of a function is any set which contains all the answers. In this case, {−6, 5, . . ., 7, 8} was used. The set of answers is called the *range* of the function. Hence, the range = {−3, −1, 1, 3, 5}. The value that any given element of the domain is mapped onto is called its *image*. For example, since f(2) = 3, then 3 is the image of 2.

5.3   Composition of Functions

If $f : x \longmapsto x + 3$ and $g : x \longmapsto x^2 + 1$, then fg is called the *product* or *composition* of f and g. It is used to denote g applied first, followed by f applied to the answer. Since g(2) = 5 and f(5) = 8, it follows that fg(2) = 8.
Also, f(2) = 5 and g(5) = 26, hence gf(2) = 26.
We can see that, in general, fg ≠ gf.

Composition of functions is *not commutative*.

In general, since f adds 3 to anything,

$$f(x^2 + 1) = (x^2 + 1) + 3 = x^2 + 4,$$

hence $fg : x \longmapsto x^2 + 4$.
But, since g will square a number and add 1,

$$\begin{aligned} g(x + 3) &= (x + 3)^2 + 1 \\ &= x^2 + 6x + 9 + 1 \\ &= x^2 + 6x + 10, \end{aligned}$$

hence $gf : x \longmapsto x^2 + 6x + 10$.

5.4   Inverse

Referring back to Fig. 5.1, we could also argue that the values of *x* are related to the values of *y*. We can show this (see Fig. 5.3) simply by reversing the arrows.

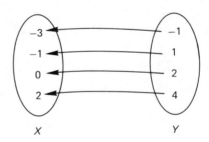

Fig. 5.3

The 'rule' is clearly to subtract 2. This rule is called the inverse of f, written f^{-1}. Hence, $f^{-1} : y \longmapsto y - 2$. It could also be written $f^{-1} : x \longmapsto x - 2$. The inverse relation can usually be found by changing the subject of the original function. See 3.16. For example, if $g : x \longmapsto 2x + 1$, then, writing this as

$$y = 2x + 1,$$
$$y - 1 = 2x,$$
$$\therefore \quad \frac{y - 1}{2} = x.$$

Hence, $g^{-1} : y \longmapsto \frac{y - 1}{2}.$

Worked Example 5.1

If $g(x)$ denotes the sum of all the prime numbers which are less than x, find (a) $g(16)$; (b) a value of x such that $g(x) = 77$; (c) the solution of the equation $g(x) = g(10)$.

Solution

(a) $g(16) = 2 + 3 + 5 + 7 + 11 + 13 = 41$.
(b) This is basically trial and error.
   Since $2 + 3 + 5 + 7 + 11 + 13 + 17 + 19 = 77$, then any number greater than 19 and less than the next prime number which is 23 works.
      Hence $x = 20$ will do.
(c) If $g(x) = g(10)$, then $x = 10$ is an obvious solution.
   However,

$$g(10) = 2 + 3 + 5 + 7 = 17, \quad \text{also} \quad g(8) = 17, g(9) = 17 \quad \text{and} \quad g(11) = 17.$$

Hence the solution is $\{8, 9, 10, 11\}$.

Worked Example 5.2

For any positive integer n, $T(n)$ is defined as the smallest multiple of 3 which is larger than or equal to n, e.g. $T(5) = 6$.
(a) Write down the value of $T(9)$ and that of $T(10)$.
(b) Give the solution set of the equation $T(n) = 12$.
(c) State whether each of the following equations is satisfied by all values of n, by some values of n, or by no values of n. Give the solution set of each equation which is satisfied by some values of n.
   (i) $T(n + 1) = T(n) + 1$; (ii) $T(n + 3) = T(n) + 3$; (iii) $T(2n) = 2T(n)$.   [L]

Solution

This example illustrates a function which cannot easily be represented by formulae.
(a) $T(9) = 9$, $T(10) = 12$.
(b) **Try to express in words what this says**: the smallest multiple of 3 which is larger than or equal to n is 12. Hence $n = 10$, 11 or 12. The solution set is 10, 11, 12.
(c) (i) Since $T(n)$ is a multiple of 3, $T(n) + 1$ is not a multiple of 3, therefore it cannot be $T(n + 1)$. Therefore (i) is not satisfied by any values of n.
   (ii) This is true for all values of n. (Convince yourself with a few examples.)
   (iii) This is not always true, since $T(10) = 2T(5)$ but $T(14) \neq 2T(7)$.
      It is true, however, for $n \in \{1, 4, 7, 10, \ldots\}$.

5.5   Graphical Representation

The reader should have some knowledge of graph plotting. A graph can be used to represent a function or mapping. The values of the domain and function are tabulated in a *table of values*.
(a) Consider the function $f : x \longmapsto 3x - 2$ for the domain $\{x \in \mathbb{R} : -3 \leqslant x \leqslant 3\}$.

The following is the table of values:

x	-3	-2	-1	0	1	2	3	Domain
$y = 3x - 2$	-11	-8	-5	-2	1	4	7	Image (range)

These points can now be plotted on a graph. If all values of the domain had been used, a continuous straight line would have resulted: see Fig. 5.4.

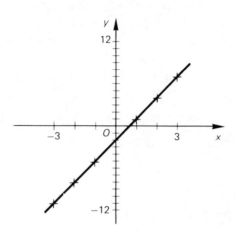

Fig. 5.4

Note that by changing the scale for the y-values, a better-shaped graph is created.

(b) Now a more complicated example. Consider the function

$$f : x \longmapsto x^3 + x^2 - 4x - 4$$

for the domain $\{x : -3 \leqslant x \leqslant 3\}$.

It is often helpful to split the function up into different parts when forming the table.

This has been done in Table 5.1.

Table 5.1

x			-3	-2	-1	0	1	2	3
x^3			-27	-8	-1	0	1	8	27
	x^2		9	4	1	0	1	4	9
		$-4x$	12	8	4	0	-4	-8	-12
		-4	-4	-4	-4	-4	-4	-4	-4
$y = x^3 + x^2 - 4x - 4$			-10	0	0	-4	-6	0	20

The graph is shown plotted in Fig. 5.5. Note that between $x = -2$ and $x = -1$, the curve must not be drawn flat. To be sure, extra values of x should be tried. If $x = -1.5$, $y = 0.875$. The trough of the curve is in fact at $x = 0.87$.

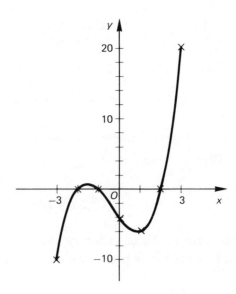

Fig. 5.5

5.6 The Equation of a Straight Line

(a) The simplest form of the equation of a straight line is $y = mx + c$. When $x = 0$, $y = c$.

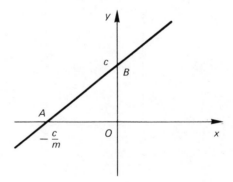

Fig. 5.6

When $y = 0$, $\qquad\qquad 0 = mx + c$

$$\therefore \quad mx = -c$$

$$\therefore \quad x = -\frac{c}{m}.$$

The gradient of the line is $\dfrac{OB}{OA}$; see Fig. 5.6.

Hence gradient $= \dfrac{c}{\dfrac{c}{m}} = m$.

(b) Other forms:

$$ax + by + c = 0, \qquad\qquad (1)$$

$$\frac{x}{a} + \frac{y}{b} = 1. \qquad\qquad (2)$$

(1) Consider the equation $3x + 5y + 4 = 0$;
make y the subject of this formula;

$$\therefore \ 5y = -3x - 4.$$

$\div 5: \quad y = -\dfrac{3x}{5} - \dfrac{4}{5}.$

Hence the gradient of the line is $-\frac{3}{5}$, and it cuts the y-axis at $(0, -\frac{4}{5})$.

(2) Put $x = 0$, $\dfrac{y}{b} = 1$; $\therefore \ y = b$.

Put $y = 0$, $\dfrac{x}{a} = 1$; $\therefore \ x = a$.

This form of the equation gives the points where it cuts the axes straight away. It can be particularly useful in linear programming problems.

5.7   Simple Coordinate Geometry (Distance, Mid-point)

If we want to find the distance between two points $P\,(x_1, y_1)$ and $Q\,(x_2, y_2)$, then a simple right-angled triangle can be constructed with sides parallel to the axes (see Fig. 5.7).

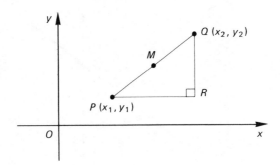

Fig. 5.7

$$PQ = \sqrt{QR^2 + PR^2}$$
$$= \sqrt{(y_1 - y_2)^2 + (x_1 - x_2)^2}.$$

The mid-point M of PQ is a simple average of the coordinates of P and Q.

$$\therefore \ M \text{ is } \left(\frac{x_1 + x_2}{2}, \frac{y_1 + y_2}{2}\right).$$

5.8   Equation of a Circle

Consider any point $P\,(x, y)$ on the circle centre O; see Fig. 5.8.

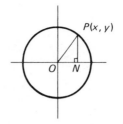

Fig. 5.8

By Pythagoras,

$$OP^2 = ON^2 + NP^2 = x^2 + y^2.$$

But OP will always be the radius of the circle r,

Hence $x^2 + y^2 = r^2.$

This is the *simplest* form of the equation of a circle.

5.9 Graphical Solution of Equations

In section 3.14, it was stated that simultaneous equations can be solved graphically. For example, to solve $3x + 2y = 5$ and $4x - 2y = 7$, simply plot the two straight lines in the form $y = \dfrac{-3x}{2} + \dfrac{5}{2}$ and $y = 2x - \dfrac{7}{2}$. Where the two lines intersect (x, y) gives the solution of the equations. We can, however, tackle more complicated examples.

Worked Example 5.3

Solve by a graphical method the equation $x^3 = 2x + 1$. Without too much extra work, how could you solve the equation $2x^3 - 3 = 4x$?

Solution

To solve $x^3 = 2x + 1$ is fairly easy. If you plot $y = x^3$ and $y = 2x + 1$, when the lines cross, the two y values are equal, hence $x^3 = 2x + 1$. The values of x where the curves cross will be the solution of the equation.

   The most difficult part of this is plotting $y = x^3$. It is worth seeing if this can be used again.

Consider $2x^3 - 3 = 4x$;

$$2x^3 = 4x + 3; \therefore x^3 = 2x + \frac{3}{2}.$$

Hence, by just drawing the straight line $y = 2x + \dfrac{3}{2}$, the second equation can be solved.

Worked Example 5.4

(a) Copy and complete Table 5.2, calculating, for values of x from 1 to 6, the value of y, where

$$y = 2 + \frac{4}{x^2}.$$

Table 5.2

x	1	1.5	2	2.5	3	4	5	6
x^2		2.25		6.25	9			36
$\dfrac{4}{x^2}$		1.78		0.64	0.44			0.11
$y = 2 + \dfrac{4}{x^2}$		3.78		2.64	2.44			2.11

(b) Using a scale of 2 cm to represent one unit on the x-axis and 4 cm to represent one unit on the y-axis, draw the graph of the curve whose equation is $y = 2 + \dfrac{4}{x^2}$, for values of x from $x = 1$ to $x = 6$.

(c) Using the same axes and the same scales, draw a graph of the straight line whose equation is $y = \frac{1}{3}x + 1$.

(d) Write down, but do not simplify, the equation in x satisfied by the value of x at the point of intersection of the curve and the straight line.

(e) Show that the equation you have written in (d) can be simplified to $x^3 - 3x^2 - 12 = 0$.

(f) Read off, from your graphs, an approximate solution of the equation in (e).

[AEB]

Solution

This question is slightly more difficult numerically but this should not deter you.

(a) Table 5.3 shows only the *missing values:*

Table 5.3

x	1	1	4	5
x^2	1	4	16	25
$\dfrac{4}{x^2}$	4	1	0.25	0.16
y	6	3	2.25	2.16

(b) The graph of the function is shown in Fig. 5.9.

(c) If you can *safely* recognize $y = \frac{1}{3}x + 1$ is a straight line, only **three** points need be plotted:

$$x = 0, y = 1; \quad x = 3, y = 2; \quad x = 6, y = 3.$$

Choose values of x to get a good spread.

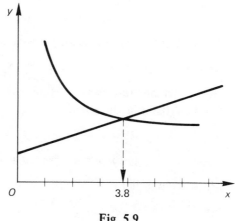

Fig. 5.9

(d) When the two graphs meet, both y-values are equal,

$$\therefore \tfrac{1}{3}x + 1 = 2 + \frac{4}{x^2} .$$

(e) Multiply by $3x^2$:

$$3x^2 \times \frac{1}{3}x + 3x^2 \times 1 = 3x^2 \times 2 + 3x^2 \times \frac{4}{x^2}$$

$$\therefore \quad x^3 + 3x^2 = 6x^2 + 12.$$

Collect terms on left-hand side:

$$x^3 - 3x^2 - 12 = 0.$$

(f) The value of x where the two lines meet is $x = 3.8$.

Worked Example 5.5

(a) The point A (2, 5) is mapped onto the point B (2, -4) under reflection in the line $y = mx + c$. Find m and c.

(b) The lines $2y - x = 8$ and $y + mx = 4$ are parallel; find m.

(c) Find the coordinates of the point C (3, 4) after a rotation of 180° about the point D (5, -1).

Solution

A reasonably accurate diagram is often helpful in questions like this.

(a) Since the x-coordinate is unchanged, the mirror line must be parallel to the y-axis.

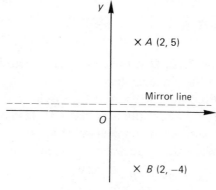

Fig. 5.10

$$\therefore\ m=0,\ \therefore\ \text{line is } y = c.$$

The distance between A and B is 9. Mirror line is halfway.

$$\therefore\ c = 5 - 4\tfrac{1}{2} = \tfrac{1}{2}.$$

(b) **Questions on gradients of lines are often much easier if the equation of the line is written in the form** $y = mx + c$.

$$2y - x = 8 \quad \Rightarrow \quad 2y = x + 8 \quad \Rightarrow \quad y = \tfrac{1}{2}x + 4;$$

$$y + mx = 4 \quad \Rightarrow \quad y = -mx + 4.$$

Lines are parallel if the gradients are equal,

$$\therefore\ \tfrac{1}{2} = -m, \quad \therefore\ m = -\tfrac{1}{2}.$$

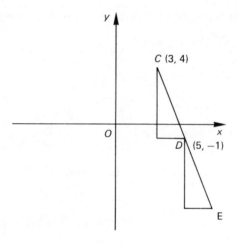

Fig. 5.11

Since CD rotates by $180°$, CDE is a straight line, and $CD = DE$.

$$\therefore\ \overrightarrow{CD} = \begin{pmatrix} 2 \\ -5 \end{pmatrix},$$

$$\text{hence } \overrightarrow{DE} = \begin{pmatrix} 2 \\ -5 \end{pmatrix}.$$

Starting at D $(5, -1)$ and using the translation $\begin{pmatrix} 2 \\ -5 \end{pmatrix}$, we find that E is $(7, -6)$.

Worked Example 5.6

In Fig. 5.12, O is the centre of the circle which passes through A $(-6, 0)$. B and P lie on the circle as shown.

(a) Find the equation of the circle, and the equation of the line AB.

(b) If P has coordinates $(k, -3)$, find the exact value of k, leaving your answer as a surd in its simplest form.

(c) Write down the size of angle APB, giving a reason for your answer, and prove that $k \cos A\hat{P}B = \dfrac{3\sqrt{6}}{2}$.

[SEB]

94

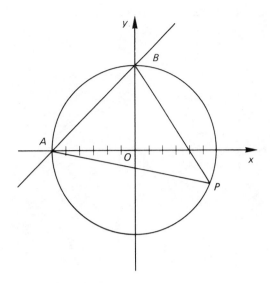

Fig. 5.12

Solution

(a) The radius of the circle is 6,
∴ equation of circle $= x^2 + y^2 = 36$.
The gradient of AB is 1, and it passes through $(0, 6)$,
∴ equation of AB is $y = x + 6$.

(b) Since P lies on the circle, $(k, -3)$ must satisfy the equation,

$$\therefore\ k^2 + (-3)^2 = 36,$$

$$\therefore\ k^2 = 27,$$

$$\therefore\ k = \sqrt{27}\ \text{(since } k > 0).$$

$$\text{i.e. } k = 3\sqrt{3}.$$

(c) $\angle APB = 45°$, because from the circle properties $\angle AOB = 2\angle APB$; see section 10.8.

$$\therefore \cos A\hat{P}B\ = \cos 45° = \frac{1}{\sqrt{2}}\,,$$

$$\therefore k \cos A\hat{P}B = \frac{1}{\sqrt{2}} \times 3\sqrt{3} = \frac{3\sqrt{3}}{\sqrt{2}} = \frac{3\sqrt{3} \cdot \sqrt{2}}{\sqrt{2} \cdot \sqrt{2}}$$

$$= \frac{3\sqrt{6}}{2}\,.$$

5.10 Curve Sketching

It is often useful to have some idea of what a curve looks like without accurate plotting. We have looked at the straight line. Other types are shown in Fig. 5.13.

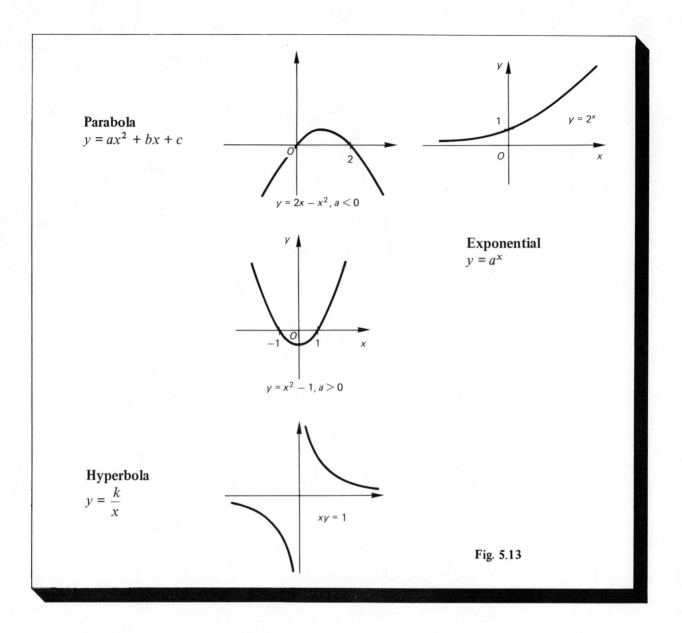

Parabola
$y = ax^2 + bx + c$

$y = 2x - x^2, a < 0$

$y = x^2 - 1, a > 0$

Exponential
$y = a^x$

$y = 2^x$

Hyperbola
$y = \dfrac{k}{x}$

$xy = 1$

Fig. 5.13

5.11 Variation and Proportion

If y is proportional to x, the graph of y against x will be a straight line.
Conversion graphs are examples.

Worked Example 5.7

A cylinder of height h and radius r has volume V. Sketch the graph of the variation in V with r, if h remains constant.

Solution

Always write down an equation which describes the problem.

In this case $V = \pi r^2 h$.
Replace any quantities by k which are not varying.
$\therefore V = kr^2$.

The graph will look like that shown in Fig. 5.14. Note that you cannot have a negative radius.

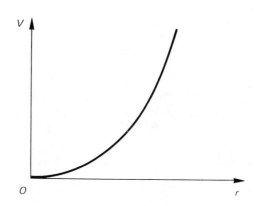

Fig. 5.14

Check 22

1 State the range of the following functions:
  (a) f : $x \longmapsto 2x - 1, \{-3 \leqslant x \leqslant 3\}$
  (b) g : $x \longmapsto x^2, x \in \{-3, -2, -1, 0, 1\}$
  (c) h : $x \longmapsto \dfrac{1}{x - 1}, x \in \mathbb{R}, x \neq 1$
  (d) m : $x \longmapsto \sin x°, \{0 \leqslant x \leqslant 360\}$

2 If f : $x \longmapsto 2x + 1$, g : $x \longmapsto x^2$ and h : $x \longmapsto 2x - 3$, find:
  (a) fg   (b) gh   (c) h²   (d) (fg)h

3 If $f(x) = \dfrac{2x + 1}{3x}$, find:
  (a) f(1)   (b) f($-\frac{1}{2}$)   (c) f(4)

4 Given that f : $x \longmapsto x^2 - 14x + 33$, find:
  (a) x if $f(x) = 0$
  (b) a value of x such that $f(x) < 0$

5 $G(x)$ denotes the sum of all positive integers less than x which are perfect squares. Find:
  (a) G(25)   (b) G(26)

6 The function g is defined on the domain $\{-2, -1, 0, 1, 2, 3\}$ by

$$g : x \longmapsto \text{units digit of } x^3.$$

Draw a mapping diagram for this function.

7 Plot the function f : $x \longmapsto x^3 - 2x^2 + 1$ for the domain $\{-3 \leqslant x \leqslant 3\}$.

8 Choose a suitable domain in order that the range of the function

$$g : x \longmapsto 2x^2 - 4$$

should not be outside the limits of -6 to 10. Using this domain, plot the graph. What is the minimum value of $g(x)$ for this domain?

9 Find the equations of the straight lines joining the following pairs of points:
  (a) (0, 4) and (3, 6)   (b) (0, −2) and (1, 1)
  (c) (2, 0) and (0, −3)   (d) (3, 1) and (2, 4)

10 Find the mid-points of the join of the pairs of points given in question 9.

Exercise 5

1 If $f : x \longmapsto -2x$ and $g : x \longmapsto x - 1$,
then $gf^{-1}(2) =$
A 1   B 2   C −1   D −2

2 The domain of the function $f : x \longmapsto \dfrac{1 - x^2}{1 + x}$ is $0 \leqslant x \leqslant \frac{1}{2}$.

The range is:
A $1 \leqslant y \leqslant 2$ B $\frac{1}{2} \leqslant y \leqslant 1$ C $0 \leqslant y \leqslant \frac{3}{4}$ D $\frac{3}{4} \leqslant y \leqslant 1$

3 The area enclosed by the line $2x - 5y + 20 = 0$ and the x- and y-axes is:
A 40   B 20   C 10   D none of these

4 Given that $f : x \longmapsto \dfrac{1 - x}{1 + x}$, then f^{-1} can be expressed as:

A $x \longmapsto \dfrac{1 - x}{1 + x}$ B $x \longmapsto \dfrac{1 + x}{1 - x}$ C $x \longmapsto \dfrac{x + 1}{x - 1}$ D $x \longmapsto \dfrac{x - 1}{x + 1}$

5 The equation of the line through the point $(4, 3)$ parallel to $y = 2x - 7$ is:
A $y = 2x + 3$ B $y = 2x - 4$ C $y = 2x - 5$ D $y = 2x + 5$

6 Which line is perpendicular to $2y = 3x - 7$?
A $2y + 3x + 5 = 0$ B $2y - 3x + 5 = 0$ C $3y - 2x + 5 = 0$
D $3y + 2x - 5 = 0$

7 The gradient of the line $\dfrac{x}{5} + \dfrac{y}{3} = 1$ is:

A $\frac{5}{3}$ B $-\frac{3}{5}$ C $\frac{3}{5}$ D $-\frac{5}{3}$

8 The line $y = 2x - 3$ passes through the point:
A $(3, 0)$ B $(0, 3)$ C $(-3, 0)$ D $(0, -3)$ [AEB]

9 A function f is defined by $f : x \longmapsto x^2 - 2$.
If $f(t) = t$, then t equals:
A 1   B 0   C −2 or 1   D 2 or −1

10 Given that $f(x) = \dfrac{k}{x^2} + 1, x \neq 0$, and $f(3) = 5$,
then k equals:
A $\frac{2}{3}$ B $\frac{14}{9}$ C 24   D 36   E 44   [SEB]

11 The quadratic function whose graph is symmetrical about $x = 0$ is:
A $f(x) = (x - 3)^2 - 4$ B $f(x) = (x - 3)^2$
C $f(x) = x^2 - 3$ D $f(x) = (x - 2)(x + 3)$

12 Draw sketches to illustrate the following equations:

(a) $y = 2x^2$; (b) $y = x^3 - 2$; (c) $y = 3^x$; (d) $y = \dfrac{-4}{x^2}$.

13 A cylinder of variable radius r has constant height h. Its volume is denoted by V. Draw a sketch graph to show the relationship between V and r as r varies.

14 Given that $f : x \longmapsto \log x$ for all positive values of x,
(a) find the value of $f(0.18)$ correct to two decimal places;
(b) find the value of x, correct to two decimal places, such that $f(x) = -0.18$.

15 The line L has equation $2x - y + 6 = 0$ and intersects the x-axis at P and y-axis at Q. The midpoint of PQ is M.
Find:
(a) the gradient of L; (b) the coordinates of M; (c) the coordinates of the image of M under an enlargement with centre the origin and scale factor −2.   [UCLES]

16 (a) (i) Write down an expression for the gradient of the line joining the points $(6, k)$ and $(4, 1)$.
(ii) Find the value of k if this gradient is $\frac{3}{5}$.
(b) Find the equation of the line through the point $(-4, 5)$ with gradient −2.

17 (a) Given that $y = 2x^2 (1.6 + x)$, copy and complete the following table:

x	-3	-2.5	-2	-1.5	-1	-0.5	0	1
$y = 2x^2 (1.6 + x)$	-25.2		-3.2					5.2

    (b) Using suitable scales to use as much of a sheet of graph paper as possible, plot the graph for x between -3 and 1 inclusive.

    (c) On the same axes, plot the graph of the straight line $y + 2x + 2 = 0$.

    (d) Write down the equation in the form $f(x) = 0$ whose solutions are the values of x where your two lines meet. Use your graph to solve $f(x) = 0$.

18 Prepare a table of values of $y = x + \dfrac{5}{x}$ for the values $x = 1, 1.5, 2, 2.5, 3, 3.5,$ $4, 4.5, 5$. Draw the graph over this range using a scale of 4 cm for 1 unit on both axes. From your graph determine, in the range $1 \leqslant x \leqslant 5$,

    (a) the least values of $x + 5/x$ and the value of x at which this occurs;

    (b) two solutions of the equation $x^2 - 4.8x + 5 = 0$.   [OLE]

19 (a) Copy and complete the following table of values of $y = 2^{-x}$ for values of x from -4.5 to 2:

x	-4.5	-4	-3	-2	-1	0	1	2
y	22.6		8				0.5	

    (b) Using a scale of 2 cm to represent one unit on the horizontal x-axis and two units on the vertical y-axis, draw the graph of $y = 2^{-x}$.

    (c) From your completed graph estimate:

       (i) the value of $2^{-0.3}$;

      (ii) the value of x for which $2^{-x} = 17$;

      (iii) the square root of 2^7, showing clearly on the diagram how your answer was obtained.

    (d) Draw on your diagram the straight line which enables you to solve the equation

$$2^{-x} - 2x = 10.$$

    Hence find the solution to this equation.

    (e) The solutions to the equation $2^{-x} = a + bx$, where a and b are constants, are 0 and -3. Determine the values of a and b.

20 Use a graphical method to solve the equations $2x + 5y = 8$, $3x - 2y = 6$. Give your answers correct to 2 d.p. Use a sketch to find the range of values that need to be plotted.

21 A function f is defined by the formula

$$f(x) = 6 \tan (x + 15)°, \text{ where } x \text{ is a real number.}$$

    (a) Evaluate $f(30)$.

    (b) Write down the exact values of $f(45)$ and $f(15)$ in terms of surds. Without using approximations, show that

$$f(45) - f(15) = 4\sqrt{3}.$$

    (c) Find the possible values of k between 0 and 360 such that $f(k) = 3$.

                                                        [SEB]

22 The line $2y + x = 10$ cuts the axes of x and y at A and B respectively. Find:

    (a) the coordinates of the points A and B;

    (b) the gradient of AB;

    (c) the equation of the line L which passes through the origin, O, and is perpendicular to AB;

(d) the coordinates of the point of intersection of the line L with AB;

(e) the coordinates of the point D, given that the quadrilateral $OBDA$ is a kite;

(f) the coordinates of the centre of the circle which passes through the points O, B, D and A;

(g) the radius of this circle. [NISEC]

23 A function f is defined by $\mathrm{f} : x \longmapsto \dfrac{x}{4} + \dfrac{4}{x} - 3$ for all values of x, except $x = 0$. $A\,(3, a)$ and $B\,(5, b)$ are points on the graph of the function f.

(a) Calculate the values of a and b and also the gradient of the chord AB.

(b) Calculate exactly the gradient m of the curve at $x = 4$. Find also the co-ordinates of the other point on the curve at which the gradient has the same value m. See section 15.1.

(c) Estimate the area between the graph of f, the x-axis and the lines $x = 3$ and $x = 5$. Use the trapezoidal rule with strips 1 unit in width. [AEB]

24 The functions f, g and h map the set of all rational numbers to itself and

$$\mathrm{f} : x \longmapsto 2, \quad \mathrm{g} : x \longmapsto \frac{x}{3} + 1, \quad \mathrm{h} : x \longmapsto x^2.$$

(a) Write down a formula for the mapping obtained:
(i) by using g followed by f; and (ii) by using f followed by h.

(b) Express the function $x \longmapsto \dfrac{x^2 + 3}{3}$ (with domain the same as f, g and h) in terms of f, g and h.

(c) Find a function m such that $\mathrm{m}(\mathrm{g}(x)) = x$.

25 $\mathrm{R}(x)$ denotes the remainder when x is divided by 3. Find:
(a) $\mathrm{R}(20)$ (b) $\mathrm{R}(20m)$ (c) $\mathrm{R}(21n)$
where m and n are integers.

26 Find the inverse of the following functions:
(a) $\mathrm{f} : x \longmapsto 3 - 2x$ (b) $\mathrm{g} : x \longmapsto \dfrac{2}{3x}$

(c) $\mathrm{h} : x \longmapsto x^3$ (d) $\mathrm{m} : x \longmapsto 2 + x^3$

(e) $\mathrm{n} : x \longmapsto \dfrac{1}{2x + 1}$ (f) $\mathrm{q} : x \longmapsto \dfrac{ax + b}{cx + d}$

27 Using the domain $-4 \leqslant x \leqslant 4$, plot the graphs of the following functions. In each case:

(a) state the greatest and least values of the function in the range;

(b) by drawing a suitable straight line, solve if possible the equation $\mathrm{f}(x) = x - 1$, giving the equation that you have solved in its simplest form.
(i) $\mathrm{f} : x \longmapsto x^2$ (ii) $\mathrm{f} : x \longmapsto 2x^2 - 3$

(iii) $\mathrm{f} : x \longmapsto \dfrac{1 + x}{x}$ $(x \neq 0)$ (iv) $\mathrm{f} : x \longmapsto x^3$

(v) $\mathrm{f} : x \longmapsto 1 - \dfrac{x^2}{x - 5}$

28 $|x|$ denotes the numerical value of x, e.g. $|-4| = 4$, $|8| = 8$.

On the same diagram, for values of x between -3 and 3 inclusive, plot the graphs of the functions:

$$\mathrm{f} : x \longmapsto |x^2 - 4| \quad \text{and} \quad \mathrm{g} : x \longmapsto 3.$$

How many solutions are there to the equation $|x^2 - 4| = 3$ for this domain? Read off the largest value as accurately as possible.

29 The function f maps x onto $\mathrm{f}(x)$ where $\mathrm{f}(x) = x^2 - 5x - 36$ and the function g maps x onto $\mathrm{g}(x)$ where $\mathrm{g}(x) = 3x - 17$. Find:
(a) $\mathrm{f}(-2)$ (b) $\mathrm{g}(5)$ (c) $\mathrm{f}(\mathrm{g}(5))$

(d) the values of x if f maps x onto 0

(e) the values of x, correct to two decimal places, if $f(x) = g(x)$.   [WJEC]

30 The function f is defined on the domain of real numbers by $f : x \longmapsto x^2 + 6$.

(a) Calculate $f(-3.58)$, giving your answer correct to three significant figures.

(b) Calculate two values of x such that $f(x) = 34$, giving your answers correct to three significant figures.

(c) Solve the equation $f(2x) = f(2x + 7)$.

(d) Sketch the graph of f.

(e) State the range of f.

The operation $*$ is defined on the set of real numbers by $a * b = f(a - b)$, where f is the function defined above.

(f) Give a numerical example to illustrate that the operation $*$ is commutative.

(g) Give a numerical example to illustrate that the operation $*$ is not associative.   [JMB]

31 Using a scale of 2 cm to 1 unit on each axis, draw the graph of $y = x - \dfrac{2}{x}$ for

values of x from 1 to 6.

Using the same axes and scales, draw the graph of $y = \frac{2}{3}(5 - x)$. From your graphs, find:

(a) one root of the equation $x - \dfrac{2}{x} = \frac{2}{3}(5 - x)$;

(b) one root of the equation $x - \dfrac{2}{x} = 3$;

(c) the set of values of x for which

$$0 < \left(x - \frac{2}{x}\right) - \tfrac{2}{3}(5 - x) < 3,$$

giving your answer in the form $a < x < b$, where the values of a and b are to be found.   [O & C]

6 Inequalities

6.1 Properties of $<$ and $\leqslant$

$a < b$ means a is less than b or b is greater than a.
$a \leqslant b$ means a is less than or equal to b, etc.
(a) Addition or subtraction of a constant:
$$a < b \quad \Rightarrow \quad a \pm k < b \pm k.$$
(b) Multiplication or division by a constant:
If $k > 0, a < b \quad \Rightarrow \quad ka < kb$,
$\quad k < 0, a < b \quad \Rightarrow \quad ka > kb.$ ⟵——— [Note that the inequality sign is reversed.]
These rules will be demonstrated by the following examples.

6.2 Inequations

(a) Solve $x + 5 < 9$.
First, subtract 5 from each side: $\qquad x < 4$.
There are an infinite number of solutions; the *solution set* is $\{x : x < 4\}$.

(b) Solve $2x - 3 \geqslant 8$.
Add 3 to each side: $2x \geqslant 11$.
Divide by $+2$. $\qquad x \geqslant 5\frac{1}{2}$.

(c) Solve $3(x + 4) < 5x + 9$.
$3x + 12 < 5x + 9$.
Subtract 12 from each side: $3x < 5x - 3$.
Subtract $5x$ from each side: $-2x < -3$.
Divide by -2 $x > \frac{3}{2}$;
note that the sign has changed.

(d) Solve $\dfrac{4}{x} > 3$.

We do not know whether x is positive or negative; therefore you have to be careful when multiplying each side by x.
If $x > 0, 4 > 3x$.
Divide by 3: $\frac{4}{3} > x$.
Since $x > 0$ the solution is $0 < x < \frac{4}{3}$.
If $x < 0, 4 < 3x$.
Divide by 3: $\frac{4}{3} < x$.
This is impossible because x cannot be both negative and greater than $\frac{4}{3}$.
The only solution is $\{x : 0 < x < \frac{4}{3}\}$.

(e) Solve $3 < 2x - 10 < 7$, given that x is an integer.

A problem of this type is best solved by splitting it into two inequations.

The left-hand side is: $3 < 2x - 10$;

$$\therefore \; 13 < 2x;$$
$$\therefore \; 6\tfrac{1}{2} < x.$$

The right-hand side is: $2x - 10 < 7$;

$$\therefore \; 2x < 17;$$
$$\therefore \; x < 8\tfrac{1}{2};$$
$$\therefore \; 6\tfrac{1}{2} < x < 8\tfrac{1}{2}.$$

But x is an integer, $\qquad \therefore \quad x \in \{7, 8\}$.

(f) Solve the quadratic inequation $x^2 - 5x + 4 > 0$.

Factorizing gives $(x - 1)(x - 4) > 0$.

Look at the number line in Fig. 6.1 divided into three regions by the points $x = 1$ and $x = 4$.

Fig. 6.1

Take any value in region I, $x = -2$;

substitute into $x^2 - 5x + 4 = (-2)^2 - 5(-2) + 4 = 18 > 0$.

Take any value in region II, $x = 3$;

substitute into $x^2 - 5x + 4 = 3^2 - 5(3) + 4 = -2 < 0$.

Take any value in region III, $x = 6$;

substitute into $x^2 - 5x + 4 = 6^2 - 5(6) + 4 = 10 > 0$.

Since we want $x^2 - 5x + 4$ to be positive, this happens in regions I and III.

The solution can be written: $\{x : x < 1 \text{ and } x > 4\}$.

Check 23

Find the solution sets of the following inequations:

1   $18 - x > 2x + 7$.

2   $3 - 2x < 4 - 3x$.

3   $3x - 5 \leqslant x + 15$ (x is a positive integer).

4   $3 - x < 14 + 5x$ (x is a negative integer).

5   $4 < x + 6 < 18$.

6   $3 < 2(x - 5) < 7$ (x an integer)

7   $-14 < 3x - 5 < 7$.

8   $-3 < 5(x + 4) < 9$.

9   $-2 \leqslant 5 - x \leqslant 3$ (x is an integer).

10   $x^2 - 3x - 4 < 0$.

11   $x^2 - 5x + 3 < -3$.

12   $x^2 - 12 < 3$.

13   $6 - x - x^2 > 0$.

14   $\dfrac{2}{5} < \dfrac{1}{x}$.

15   $\dfrac{1}{2 - x} < \tfrac{1}{4}$ ($x \neq 2$).

6.3   Errors

v, u, a and t are related by the equation $v = u + at$. If u, a and t cannot be measured exactly, then there is uncertainty in the value of v. If u, a and t can be guaranteed to the nearest whole number, what can be said about v when $u = 40$, $a = 4$ and $t = 15$? Since the values are only correct to the nearest whole number, then $39.5 \leqslant u \leqslant 40.5$, $3.5 \leqslant a \leqslant 4.5$ and $14.5 \leqslant t \leqslant 15.5$.

Although 40.5 would be rounded up to 41, it is included to keep the working simple.

If the smallest values of u, a and t are taken, then $v = 39.5 + 3.5 \times 14.5 = 90.25$.
If the largest values of u, a and t are taken, then $v = 40.5 + 4.5 \times 15.5 = 110.25$.
Hence $90.25 \leqslant v \leqslant 110.25$.
Clearly, if this were a scientific experiment, we could be in trouble.

Worked Example 6.1

It is given that $H = \dfrac{4a}{2 + 4t}$, with a and t measured in centimetres to the nearest centimetre. If $a = 6$ and $t = 4$, find the range of possible values of H.

Solution

Since a and t are measured to the nearest whole number, a can be anywhere between 5.5 and 6.5,

i.e. $\qquad\qquad\qquad\qquad\qquad\qquad\quad 5.5 \leqslant a \leqslant 6.5$.
Similarly, $\qquad\qquad\qquad\qquad\qquad\quad 3.5 \leqslant t \leqslant 4.5$.

Strictly speaking, we should exclude 6.5, as this would normally be rounded up. But this leads to complications, and does not really affect the result. Since H is a fraction, the largest value of H will have the largest numerator, and the smallest denominator. The smallest value of H will have the smallest numerator and the largest denominator.

Hence $\qquad\qquad\qquad \dfrac{4 \times 5.5}{2 + 4 \times 4.5} \leqslant H \leqslant \dfrac{4 \times 6.5}{2 + 4 \times 3.5}$

and hence $\qquad\qquad\qquad\qquad 1.1 \leqslant H \leqslant 1.625$.

6.4 Regions

We can use set notation to describe regions in the Cartesian plane in quite a neat way. Consider the following example:

Worked Example 6.2

Draw a sketch to illustrate the following regions:
(a) $A = \{(x, y) : y \geqslant 1, x + y \leqslant 5, 3y - 2x < 6\}$;
(b) $B = \{(x, y) : y \geqslant x^2, y \leqslant x + 2\}$.

Solution

(a) We will generally operate the convention of shading the *unwanted* region. Step 1 is to plot the lines without the inequalities (see Fig. 6.2);

i.e. $\qquad\qquad\qquad y = 1, \qquad x + y = 5, \qquad 3y - 2x = 6$.

The last will be drawn dotted because the inequality does not include the equal part.
$y \geqslant 1$ means shade below $y = 1$.

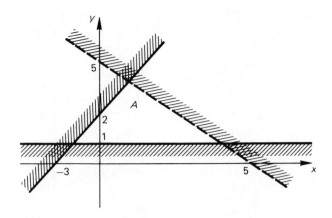

Fig. 6.2

$x + y \leqslant 5$ means shade above line $x + y = 5$.

$3y - 2x < 6$:   it is easier to rewrite this $y < \dfrac{2x}{3} + 2$ (careful with the sign);

∴ shade above this line.

(b) Step 1: plot $y = x^2$, $y = x + 2$.

$y \geqslant x^2$, ∴ shade below the curve.

$y \leqslant x + 2$, ∴ shade above the line.

The region is shown in Fig. 6.3.

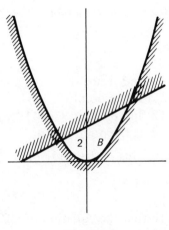

Fig. 6.3

6.5   Linear Programming

The techniques of linear programming are built around finding the maximum or minimum values of an expression $Ax + By$ in a given region. It is always worth looking at the value of $Ax + By$ at the vertices of the region first. The following worked examples should illustrate the ideas involved.

Worked Example 6.3

In Fig. 6.4 the line with equation $x + y = 12$ crosses the y-axis at A and meets the line $x = 6$ at B. C is the point $(0, 3)$ and M is the mid-point of AB.

(a) Find:

    (i) the coordinates of A, B and M;

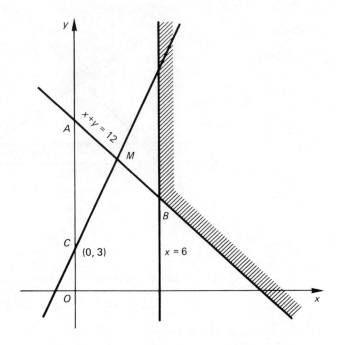

Fig. 6.4

    (ii) the equation of the line CM.

(b) If x and y are subject to the restrictions

$$y \geqslant 0, \qquad x \geqslant 6 \qquad \text{and} \qquad x + y \geqslant 12,$$

deduce the least value of $2x + 3y$.   [SEB]

Solution

(a) (i) To find A, use $x + y = 12$ with $x = 0$, i.e. $y = 12$.

        $\therefore A$ is the point $(0, 12)$.

        To find B, use $x + y = 12$ with $x = 6$, i.e. $y = 6$.

        $\therefore B$ is the point $(6, 6)$.

To find M, this is the mid-point of $(0, 12)$ and $(6, 6)$. Using $\left(\dfrac{x_1 + x_2}{2} , \dfrac{y_1 + y_2}{2} \right)$

(see section 5.7),

M is $(3, 9)$.

      (ii) To find the equation of CM, we need its gradient,

        i.e. $\dfrac{9 - 3}{3 - 0} = 2$.

        It passes through $(0, 3)$.

        $\therefore$ Equation of line is $y = 2x + 3$.

(b) To satisfy $y \geqslant 0$, $x \geqslant 6$ and $x + y \geqslant 12$, the point (x, y) lies in the shaded region. The least value of $2x + 3y$ is bound to occur at $B(6, 6)$;

    $\therefore$ least value $= 2 \times 6 + 3 \times 6 = 30$.

Worked Example 6.4

A man has 12 acres of land available for growing peas and beans. Each acre of peas that he plants costs him £70 and involves him in 18 hours of labour, while each acre of beans that he plants costs him £150 and involves him in 5 hours of labour. He has £1050 and 90 hours of labour available.

    Let x acres and y acres be the areas that he uses to grow peas and beans respectively and write down three inequalities other than $x \geqslant 0$ and $y \geqslant 0$. Using 1 cm to denote 1 acre on both the *x*- and *y*-axes, draw a graph to illustrate these inequalities.

Show that the man cannot use all the land that he has available, and use your graph to estimate the largest amount of land that he can use for growing peas and beans.   [OLE]

Solution

The three inequalities are determined by the number of acres, the number of hours, and the cost.

The number of acres cannot exceed 12,

$$\therefore x + y \leqslant 12. \tag{1}$$

The number of hours cannot exceed 90,

$$\therefore 18x + 5y \leqslant 90 \tag{2}$$

The cost cannot exceed £1050,

$$\therefore 70x + 150y \leqslant 1050.$$

Always simplify if possible; in this case, ÷ 10:

$$\therefore 7x + 15y \leqslant 105. \tag{3}$$

The region within which *x* and *y* must lie is denoted by *R* in Fig. 6.5.

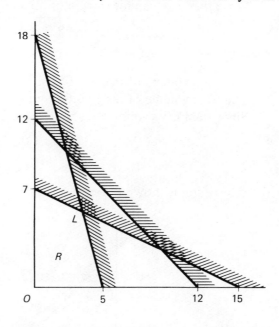

Fig. 6.5

Since the line *x* + *y* = 12 does not pass through the region *R*, there is no solution for which all the acreage is used. The largest value of *x* + *y* occurs at the point *L* (3.5, 5.4) (answers to 1 d.p.).

The largest area = 3.5 + 5.4 = 8.9 acres.

Worked Example 6.5

A depot for famine relief has 20 sacks of rice and 35 sacks of maize. The weight, volume and number of meal rations *for each sack* are as shown:

	Weight (kg)	Volume (m³)	No. of meals
Sack of			
rice	25	0.05	800
maize	10	0.05	160

A delivery van is to carry the largest possible total number of meals. It can carry up to 600 kg in weight and 2 m³ in volume.

(a) If a load is made up of x sacks of rice and y sacks of maize, say why $x \leqslant 20$ and write down three other inequalities (other than $x \geqslant 0$, $y \geqslant 0$) which govern x and y.

(b) Illustrate these inequalities on a graph and indicate the area in which the point (x, y) must lie.

(c) Write down an expression for the number of meals that can be provided from x sacks of rice and y sacks of maize. Using the graph, find the best values to take for x and y.   [NISEC]

Solution

(a) Since only 20 sacks of rice are available, we must have $x \leqslant 20$. Similarly, $y \leqslant 35$.

The total weight of rice is $25x$ kg.
The total weight of maize is $10y$ kg.
Since the lorry cannot carry more than 600 kg, it follows that

$$25x + 10y \leqslant 600.$$

$\div 5$: $$5x + 2y \leqslant 120.$$

The total volume of rice is $0.05x$ m³.
The total volume of maize is $0.05y$ m³.
Since the lorry cannot carry more than 2 m³, it follows that

$$0.05x + 0.05y \leqslant 2.$$

$\times 20$: $$x + y \leqslant 40.$$

(b) The region is shown in Fig. 6.6. The points which will be needed can be found

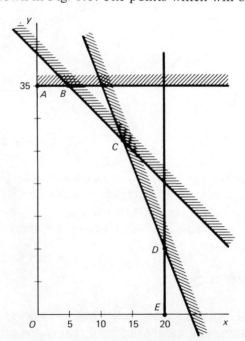

Fig. 6.6

either by drawing or by simultaneous equations. They are A (0, 35), B (5, 35), C ($\frac{40}{3}$, $\frac{80}{3}$), D (20, 10) and E (20, 0).

(c) The number of meals is

$$800x + 160y = 160 (5x + y).$$

This will be greatest when $5x + y$ is greatest. The values of $5x + y$ at the points found in (b) are A (35), B (60), C ($93\frac{1}{3}$), D (110) and E (100). The best value is at D, i.e. $x = 20$, $y = 10$.

6.6   Locus

Simple locus problems can be stated using the inequality signs; consider the following example.

Worked Example 6.6

In Fig. 6.7, $ABCD$ is a square; a point P moves inside the square so that $PA \geqslant PB$, and $PC \geqslant PA$. Shade the region in which P must lie.

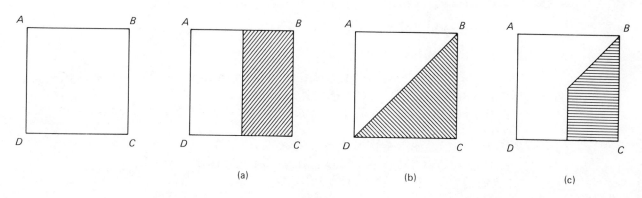

(a)   (b)   (c)

Fig. 6.7

Solution

$PA \geqslant PB$ means P is nearer to B than A. This means that P lies in the shaded region in (a).

$PC \leqslant PA$ means P is nearer to C than A. Hence P lies in the shaded region in (b). The final region is the intersection of these, shown in (c).

Exercise 6

1 Given that $p = \frac{1}{3}$, $q = \frac{2}{7}$ and $r = \frac{3}{10}$, it follows that:
      A $p < q < r$ **B** $p < r < q$ **C** $p > q > r$ **D** $p > r > q$ [AEB]

2 p is a negative integer and q is a positive integer. Which of the following is (are) true?

   I $pq > 0$ **II** $\dfrac{p}{q} < 0$ **III** $\dfrac{q}{p} < 0$ **IV** $pq < 0$

   A I and II only   **B** II and IV only   **C** II, III and IV only
   D III and IV only   **E** IV only   [SEB]

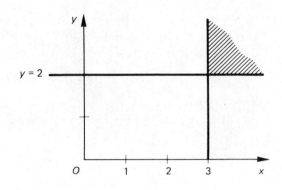

Fig. 6.8

3 The shaded area shown in Fig. 6.8 represents the set:

A $\{(x, y) : x \geqslant 3\} \cap \{(x, y) : y \geqslant 2\}$

B $\{(x, y) : x \geqslant 3\} \cup \{(x, y) : y \geqslant 2\}$

C $\{(x, y) : x \leqslant 3\} \cup \{(x, y) : y \leqslant 2\}$

D $\{(x, y) : x \leqslant 3\} \cap \{(x, y) : y \leqslant 2\}$

E none of these

4 If $S = \{(x, y) : x$ and y are positive integers and $x + y \leqslant 5\}$, then n(S) equals:
A 36 B 25 C 16 D none of these

5 If $a = 0.32$ and $b = 5.71$ are correct to two decimal places, the maximum value of $b - a$ is:
A 5.4 B 5.5 C 5.39 D 5.335

6 If $n^2 \leqslant \sqrt{300} \leqslant (n + 1)^2$ and n is an integer, then n equals:
A 17 B 18 C 3 D 4

7 If $\dfrac{1}{\sqrt{3}} < \dfrac{p}{q} < \dfrac{1}{\sqrt{2}}$, then $\dfrac{p}{q}$ could equal:
A $\frac{4}{5}$ B $\frac{1}{2}$ C $\frac{2}{3}$ D $\frac{5}{6}$

8 If $E = \dfrac{2b}{c + 2}$, and $3 \leqslant b \leqslant 7$, $2 \leqslant c \leqslant 5$, the minimum value of E is:
A 2 B $\frac{6}{7}$ C $\frac{4}{5}$ D $\frac{14}{4}$

9 If $x^a > x^b$, it follows that:
A $x > 0$ B $a > b$ C $a < b$ D none of these

10 The solution to the inequation $n^2 + n + 1 \geqslant 0$ is:
A $n \geqslant -\frac{1}{2}$ B $n \geqslant \frac{3}{4}$ C any value of n D $n \geqslant 1$

11 List the integer values of x which satisfy $3x - 4 < 27 \leqslant 4x - 5$. [UCLES]

12 Given that $1 \leqslant x \leqslant 4$ and $3 \leqslant y \leqslant 5$, find:
(a) the largest possible value of $y^2 - x$;
(b) the smallest possible value of (i) $\dfrac{y^2}{x}$; (ii) $(y - x)^2$. [UCLES]

13 A fuel merchant has two storage depots, P and Q, which hold stocks of 30 units and 20 units of fuel respectively. (A unit of fuel represents a full tanker of oil.) Three customers A, B and C place orders for 25, 15 and 10 units of fuel respectively.

Let x units of fuel be delivered from depot P to customer A and y units of fuel from depot P to customer B. Copy and complete the following table for the deliveries required to satisfy the orders:

110

	A	B	C
Stocks	25	15	10
depot P (30)	x	y	$30 - (x + y)$
depot Q (20)		$15 - y$	

(a) Assuming that no delivery may have a negative value, write down the six inequalities which must be satisfied.

(b) Illustrate the inequalities found in (a) graphically. The distances in kilometres between the depots and the customers are shown in the following table:

	A	B	C
P	5	4	2
Q	1	2	3

(c) If the cost of delivery per unit of fuel per km is £1, write down and simplify an expression in x and y for the *total* delivery cost £T. (*Hint:* The cost of delivery from P to C is $2[30 - (x + y)]$ pounds.)

(d) Use your results in (b) and (c) to determine the minimum value of T so that all the orders are fulfilled.
[AEB]

14 Mrs Jones is baking two kinds of cake for a local fête. Each cake of type A requires 150 g flour, 120 g fat and 60 g sugar, while each cake of type B requires 160 g flour, 75 g fat and 100 g sugar. She has 2.4 kg flour, 1.8 kg fat and 1.2 kg sugar available for baking the cakes.

She makes x cakes of type A and y cakes of type B. Write down three inequalities, apart from $x \geqslant 0$ and $y \geqslant 0$, which govern the possible values of x and y. Draw a graph to illustrate these inequalities, taking 1 cm to represent 1 unit on each axis. Indicate clearly the region of points whose coordinates satisfy all the conditions.

Cakes of type A sell at a profit of 25 pence each and those of type B at a profit of 40 pence each. Write down an expression, in terms of x and y, for the total profit she makes if she sells all the cakes that she bakes. What is the largest profit she could make?
[OLE]

15 A shopkeeper stocks two brands of drinks called Kula and Sundown, both of which are produced in cans of the same size. He wishes to order fresh supplies and finds that he has room for up to 1000 cans. He knows that Sundown is more popular and so proposes to order at least twice as many cans of Sundown as of Kula. He wishes, however, to have at least 100 cans of Kula and not more than 800 cans of Sundown. Taking x to be the number of cans of Kula and y to be the number of cans of Sundown which he orders, write down the four inequalities involving x and/or y which satisfy these conditions.

The point (x, y) represents x cans of Kula and y cans of Sundown. Using a scale of 1 cm to represent 100 cans on each axis, construct and indicate clearly, by shading the unwanted regions, the region in which (x, y) must lie.

The profit on a can of Kula is 3p and on a can of Sundown is 2p. Use your graph to estimate the number of cans of each that the shopkeeper should order to give the maximum profits.
[UCLES]

16 A new book is to be published in both a hardback and a paperback edition.
A bookseller agrees to purchase:
(a) 15 or more hardback copies;
(b) more than 25 paperback copies;
(c) at least 45, but fewer than 60, copies altogether.

Using h to represent the number of hardback copies and p to represent the number of paperback copies, write down the inequalities which represent these conditions.

The point (h, p) represents h hardback copies and p paperback copies. Using a scale of 2 cm to represent 10 books on each axis, construct, and indicate clearly by shading the *unwanted* regions, the region in which (h, p) must lie. Given that each hardback copy costs £5 and each paperback costs £2, calculate the number of each sort that the bookseller must buy if he is prepared to spend between £180 and £200 altogether and he has to buy each sort in packets of five. [UCLES]

7 Vectors

7.1   Definition

A vector is a quantity which has length (magnitude) and direction. A simple translation (or displacement) is a vector quantity. The notation **AB**, $\overrightarrow{AB}$, **a** will be used. In this book, a vector will be represented by a simple column vector,

e.g. $$\overrightarrow{OA} = \begin{pmatrix} 3 \\ 2 \end{pmatrix}$$

represents the vector $\overrightarrow{OA}$ shown in Fig. 7.1. Note that 3 and 2 are called the *components*.

The direction of the arrow is important: it means 'from O to A'.
$\mathbf{a} = \overrightarrow{OA}$ is the *position vector* of A with respect to O.

7.2   Multiplication by a Scalar

An ordinary number is called a *scalar*.

If $\mathbf{a} = \begin{pmatrix} 3 \\ 2 \end{pmatrix}$, then to get from O to C (Fig. 7.1), we apply **a** followed by **a**, i.e. 2**a**.

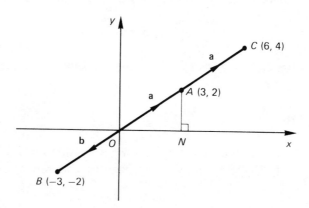

Fig. 7.1

But since $\overrightarrow{OC} = \begin{pmatrix} 6 \\ 4 \end{pmatrix}$, then $2\mathbf{a} = \begin{pmatrix} 6 \\ 4 \end{pmatrix} = 2\begin{pmatrix} 3 \\ 2 \end{pmatrix}$. Hence, to multiply a vector by a scalar, we multiply each component by the scalar.

Also, $\mathbf{b} = \begin{pmatrix} -3 \\ -2 \end{pmatrix} = -\begin{pmatrix} 3 \\ 2 \end{pmatrix} = -\mathbf{a}$.

Hence the negative of a vector is the same length in the opposite direction.

7.3 Addition and Subtraction

In Fig. 7.2,

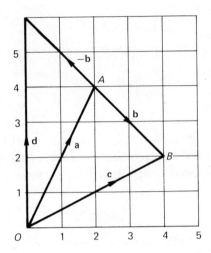

Fig. 7.2

$$\overrightarrow{OA} = \begin{pmatrix} 2 \\ 4 \end{pmatrix} = \mathbf{a}, \quad \overrightarrow{AB} = \begin{pmatrix} 2 \\ -2 \end{pmatrix} = \mathbf{b}.$$

If we start at O and apply the translation $\mathbf{a}$ followed by $\mathbf{b}$, we arrive at the point B (4, 2). We could also have arrived at B by the translation $\mathbf{c} = \begin{pmatrix} 4 \\ 2 \end{pmatrix}$.

We say that $\mathbf{c}$ is the *sum* or *resultant* of $\mathbf{a}$ and $\mathbf{b}$,

$$\therefore \mathbf{a} + \mathbf{b} = \mathbf{c}, \text{ i.e. } \begin{pmatrix} 2 \\ 4 \end{pmatrix} + \begin{pmatrix} 2 \\ -2 \end{pmatrix} = \begin{pmatrix} 4 \\ 2 \end{pmatrix}.$$

Hence to add two vectors, we simply add the components:

$$\mathbf{a} - \mathbf{b} = \mathbf{a} + (-\mathbf{b}) = \begin{pmatrix} 2 \\ 4 \end{pmatrix} + \begin{pmatrix} -2 \\ 2 \end{pmatrix} = \begin{pmatrix} 0 \\ 6 \end{pmatrix} = \mathbf{d}.$$

Hence to subtract two vectors, we subtract the components.

7.4 Length

The length or *modulus* of a vector is denoted by $|\mathbf{a}|$, $|\mathbf{OA}|$ or $|\overrightarrow{OA}|$.
In Fig. 7.1, applying Pythagoras' theorem to triangle ONA gives

$$|\mathbf{a}| = \sqrt{3^2 + 2^2} = \sqrt{13}.$$

In general, if $\mathbf{a} = \begin{pmatrix} x \\ y \end{pmatrix}$, then $|\mathbf{a}| = \sqrt{x^2 + y^2}$.

Check 24

1 If $\mathbf{a} = \begin{pmatrix} 3 \\ 1 \end{pmatrix}$, $\mathbf{b} = \begin{pmatrix} -2 \\ 1 \end{pmatrix}$ and $\mathbf{c} = \begin{pmatrix} 4 \\ -2 \end{pmatrix}$, find:

(a) $\mathbf{a} + \mathbf{b}$ (b) $\mathbf{b} - \mathbf{c}$ (c) $2\mathbf{a} - \mathbf{b}$
(d) $\mathbf{a} + \mathbf{b} + \mathbf{c}$ (e) $3\mathbf{a} - \mathbf{b} - \mathbf{c}$ (f) $|\mathbf{a}|$
(g) $|\mathbf{a} + \mathbf{b}|$ (h) $|\mathbf{a}| + |\mathbf{b}|$

2 $OABC$ is a quadrilateral $\overrightarrow{OA} = \mathbf{a}$, $\overrightarrow{AB} = \mathbf{b}$, and $\overrightarrow{BC} = \mathbf{c}$.
Write down the following vectors in terms of $\mathbf{a}$, $\mathbf{b}$ and $\mathbf{c}$:
(a) $\overrightarrow{AC}$ (b) $\overrightarrow{OC}$ (c) $\overrightarrow{OB}$

3 $RSTU$ is a rectangle. The diagonals of the rectangle meet in M. If $\overrightarrow{MR} = \mathbf{a}$ and
$\overrightarrow{MS} = \mathbf{b}$, write down in terms of $\mathbf{a}$ and $\mathbf{b}$:
(a) $\overrightarrow{RS}$ (b) $\overrightarrow{ST}$ (c) $\overrightarrow{US}$

4 X is the point $(4, 1)$, Y is the point $(-2, -1)$ and Z is the point $(5, -3)$.
Express, as a 2×1 column vector, the following:
(a) $\overrightarrow{XY}$ (b) $\overrightarrow{YZ}$ (c) $\overrightarrow{XZ}$

5 O $(0, 0)$, A $(3, 1)$ and B $(2, -1)$ are three vertices of a quadrilateral $OABC$. If
$\overrightarrow{BC} = \begin{pmatrix} 3 \\ -2 \end{pmatrix}$, what are the coordinates of C? Find $\overrightarrow{AC}$.

6 $ABCD$ is a quadrilateral, A, B, D have coordinates $(0, 2)$, $(2, 5)$, $(8, 0)$ respectively. If $\overrightarrow{AD} = 2\overrightarrow{BC}$, find the coordinates of C.
AC is produced to P so that $\overrightarrow{CP} = 2\overrightarrow{AC}$. Find the coordinates of P, and $|\overrightarrow{BP}|$.

7.5   Equations

1 If $\mathbf{a}$ and $\mathbf{b}$ are parallel vectors, then $\mathbf{a} = k\mathbf{b}$ where k is a scalar.
2 If $\mathbf{a}$ and $\mathbf{b}$ are not parallel vectors, and $h\mathbf{a} = k\mathbf{b}$, then $h = k = 0$.

Both of these results will be used in the section on geometrical proofs.
(*Note:* See Worked Example 7.2.)

7.6   Geometrical Proofs

The following three examples should indicate how vectors can be used to prove results in geometry.

Worked Example 7.1

ABC is a triangle. X is the mid-point of AB, and Y is the mid-point of BC. XY is produced to T so that $XY = YT$. Prove that $XBTC$ is a parallelogram.

Solution

Let $\overrightarrow{BX} = \mathbf{a}$, and $\overrightarrow{BY} = \mathbf{b}$. There are several other vectors that could have been used for $\mathbf{a}$ or $\mathbf{b}$. Try and choose those which give the simplest answers.

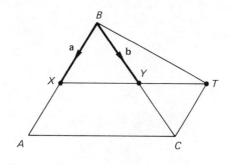

Fig. 7.3

115

$\overrightarrow{XY} = -\mathbf{a} + \mathbf{b}, \therefore \overrightarrow{XT} = 2(-\mathbf{a} + \mathbf{b}) = -2\mathbf{a} + 2\mathbf{b}.$

$\overrightarrow{BC} = 2\mathbf{b}, \therefore \overrightarrow{XC} = -\mathbf{a} + 2\mathbf{b}.$

$\overrightarrow{TC} = \overrightarrow{TX} + \overrightarrow{XC} = -\overrightarrow{XT} + \overrightarrow{XC} = 2\mathbf{a} - 2\mathbf{b} - \mathbf{a} + 2\mathbf{b} = \mathbf{a}.$

Hence $\overrightarrow{TC} = \overrightarrow{BX}$. It follows that BX is parallel to TC and the same length.

$\overrightarrow{BT} = \overrightarrow{BC} + \overrightarrow{CT} = 2\mathbf{b} - \mathbf{a} = \overrightarrow{XC}.$

Hence BT is parallel to XC and of equal length.

$\therefore XBTC$ is a parallelogram.

When trying to find an expression for, say, $\overrightarrow{XY}$, look at a route which takes you from X to Y, i.e. $X \rightarrow B \rightarrow Y$.

$\therefore \overrightarrow{XY} = \overrightarrow{XB} + \overrightarrow{BY} = -\mathbf{a} + \mathbf{b}.$

Worked Example 7.2

In the triangle OAB, $\overrightarrow{OA} = \mathbf{a}$ and $\overrightarrow{OB} = \mathbf{b}$. L is a point on the side AB. M is a point on the side OB, and OL and AM meet at S. It is given that $AS = SM$ and $OS/OL = \frac{3}{4}$; also that $OM/OB = h$ and $AL/AB = k$.

(a) express the vectors $\overrightarrow{AM}$ and $\overrightarrow{OS}$ in terms of a, b and h;

(b) express the vectors $\overrightarrow{OL}$ and $\overrightarrow{OS}$ in terms of a, b and k.

Find h and k. and hence find the values of the ratios OM/MB and AL/LB.

Solution

Since $\dfrac{OS}{OL} = \frac{3}{4}$, $\overrightarrow{OS} = \frac{3}{4}\overrightarrow{OL}$.

$\dfrac{OM}{OB} = h, \therefore OM = hOB$

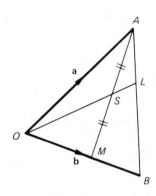

Fig. 7.4

Using vectors, therefore,

$$\overrightarrow{OM} = h\mathbf{b}$$

also

$$\overrightarrow{AL} = k\overrightarrow{AB}$$

Now

$$\overrightarrow{AB} = -\mathbf{a} + \mathbf{b} = \mathbf{b} - \mathbf{a},$$

$\therefore$

$$\overrightarrow{AL} = h(\mathbf{b} - \mathbf{a}).$$

(a) $\overrightarrow{AM} = \overrightarrow{AO} + \overrightarrow{OM} = -\mathbf{a} + h\mathbf{b}$

$\overrightarrow{OS} = \overrightarrow{OA} + \overrightarrow{AS}$

$= \overrightarrow{OA} + \frac{1}{2}\overrightarrow{AM}$

$= \mathbf{a} + \frac{1}{2}(-\mathbf{a} + h\mathbf{b}).$

Before we can use the results of section 7.5, this must be simplified.

$$\vec{OS} = \tfrac{1}{2}\mathbf{a} + \tfrac{1}{2}h\mathbf{b}.$$

(b) $\vec{OL} = \vec{OA} + \vec{AL} = \mathbf{a} + k(\mathbf{b} - \mathbf{a}) = (1 - k)\mathbf{a} + k\mathbf{b},$

$\vec{OS} = \tfrac{3}{4}\vec{OL} = \tfrac{3}{4}(1 - k)\mathbf{a} + \tfrac{3}{4}k\mathbf{b}.$

Since both expressions for $\vec{OS}$ must be the same,

$$\tfrac{1}{2}\mathbf{a} + \tfrac{1}{2}h\mathbf{b} = \tfrac{3}{4}(1 - k)\mathbf{a} + \tfrac{3}{4}k\mathbf{b},$$

$$\therefore [\tfrac{1}{2} - \tfrac{3}{4}(1 - k)]\,\mathbf{a} = (\tfrac{3}{4}k - \tfrac{1}{2}h)\,\mathbf{b}.$$

Since $\mathbf{a}$ and $\mathbf{b}$ are not parallel,

$$\tfrac{1}{2} - \tfrac{3}{4}(1 - k) = 0$$

and

$$\tfrac{3}{4}k - \tfrac{1}{2}h = 0.$$

The solutions of these are $k = \tfrac{1}{3}, h = \tfrac{1}{2}$.

$\therefore$ The ratios $\dfrac{OM}{MB} = 1, \qquad \dfrac{AL}{LB} = \tfrac{1}{2}.$

Worked Example 7.3

Y and X are points with position vectors $\mathbf{y}$ and $\mathbf{x}$ referred to an origin O. OY is produced to a point Z where $OY : YZ = 1 : 3$. V is a point on YX such that $YV : VX = 1 : 2$ and W is a point on XZ such that $XW : WZ = 1 : 2$.

Show that O, V and W lie on a straight line.

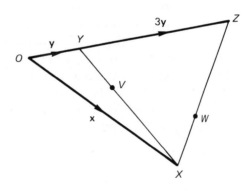

Fig. 7.5

Solution

To show that the three points lie on a straight line, try and show that $\vec{OW}$ is a multiple of $\vec{OV}$.

The positions of the points are shown in Fig. 7.5. To express $\vec{OV}$ in terms of $\mathbf{x}$ and $\mathbf{y}$, try to write $\vec{OV}$ in a different way.

$$\vec{OV} = \vec{OX} + \vec{XV} = \vec{OX} + \tfrac{2}{3}\vec{XY}$$

$$= \mathbf{x} + \tfrac{2}{3}(-\mathbf{x} + \mathbf{y}) = \tfrac{2}{3}\mathbf{y} + \tfrac{1}{3}\mathbf{x}.$$

Similarly, $\qquad \overrightarrow{OW} = \overrightarrow{OX} + \overrightarrow{XW} = x + \frac{1}{3}\overrightarrow{XZ}$

$$= x + \tfrac{1}{3}(-x + 4y)$$

$$= \tfrac{4}{3}y + \tfrac{2}{3}x.$$

Hence $\overrightarrow{OW} = 2\overrightarrow{OV}$.

It follows that OVW is a straight line, and also V is the mid-point of OW.

7.7 Locus Problems

Vector notation can often be used to represent the *locus* (path) traced out by a point which moves subject to certain conditions. The following formula is vital. If P has position vector **p**, and A has position vector a, then since **p** − **a** is the vector joining P and A, it follows that

|**p** − **a**| = **the distance between P and A**

Worked Example 7.4

If P is a variable point with position vector **r**, describe the locus of P if it moves subject to the following conditions:
(a) |**r**| = 4;
(b) |**r** − **a**| = 2 where **a** is a constant;
(c) |**r** − **a**| = |**r** − **b**| where **a** and **b** are constant;
(d) |**r** − **a**| + |**r** + **a**| = 4.

Solution

(a) If |**r**| is always 4, then the length of OP always equals 4. The locus of P is a circle, centre O, radius 4.
(b) Since |**r** − **a**| is the distance between P and A, AP always equals 2. The locus of P is a circle, centre A, radius 2.
(c) This says that the distance from P to A equals the distance from P to B. Hence the locus of P is the perpendicular bisector of AB.
(d) |**r** + **a**| can be written |**r** − (−**a**)|, which is the distance between P and a point A' with position vector −**a**. See Fig. 7.6.

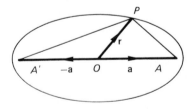

Fig. 7.6

If $AP + A'P$ always equals 4, provided |**a**| < 2, P moves round in an ellipse.

7.8 Velocity Triangles

A common problem is to find the amount a boat or aeroplane is carried off course by the wind or current. Vectors are useful in representing this. A full treatment of how vectors are used in velocity triangles is given in section 12.15.

c represents the course steered, **w** represents the wind, or current. (Normally specify the direction from which it is *coming*.) **t** represents the track it ends up following.

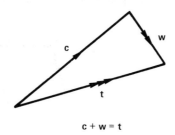

c + w = t

Fig. 7.7

Exercise 7

1 If $x = \begin{pmatrix} 3 \\ 4 \end{pmatrix}$ and $4x - 3y = \begin{pmatrix} 2 \\ -1 \end{pmatrix}$, then y equals:

   A $\begin{pmatrix} 10 \\ 17 \end{pmatrix}$ **B** $\begin{pmatrix} 14 \\ 15 \end{pmatrix}$ **C** $\begin{pmatrix} 4\frac{2}{3} \\ 5 \end{pmatrix}$ **D** $\begin{pmatrix} 3\frac{1}{3} \\ 5\frac{2}{3} \end{pmatrix}$

2 If $x = \begin{pmatrix} -4 \\ -3 \end{pmatrix}$, then $|x|$ equals:

   A -5 **B** 5   **C** 7   **D** undefined

3 If $a = \begin{pmatrix} 2 \\ 5 \end{pmatrix}$ and **b** is perpendicular to **a**, then **b** could equal:

   A $\begin{pmatrix} 2 \\ -5 \end{pmatrix}$ **B** $\begin{pmatrix} -10 \\ -4 \end{pmatrix}$ **C** $\begin{pmatrix} 10 \\ -4 \end{pmatrix}$ **D** $\begin{pmatrix} 5 \\ 2 \end{pmatrix}$

4 If **a** and **b** are non-parallel vectors, and $(2 - h)a + (3 + k)b = 0$, where h and k are numbers, then $h + k$ equals:

   A 0   **B** -1 **C** 1   **D** 5

5 *ABCDEF* is a regular hexagon. $\overrightarrow{AB} = a$ and $\overrightarrow{BC} = b$. $\overrightarrow{DF}$ equals:

   A $b - a$ **B** $a - b$ **C** $-a - b$ **D** $b + a$

6 *OPQR* is a parallelogram. $\overrightarrow{OP} = 3a + b$, $\overrightarrow{OR} = 4a + 5b$. It follows that $\overrightarrow{RP}$ equals:

   A $7a + 6b$ **B** $a + 4b$ **C** $-a - 4b$ **D** $-7a - 6b$

7 **p** and **q** are two vectors, such that **p** is perpendicular to $(p + q)$. If $|p| = 5$, and $|q| = 8$, then $|p + q|$ equals:

   A 3   **B** 13   **C** $\sqrt{39}$ **D** none of these

8 $\overrightarrow{AB} = e + 4n$, where $|e| = |n| = 1$, **e** is due east and **n** is due north. The bearing of A from B to the nearest degree is:

   A 14°   **B** 194°   **C** 76°   **D** 346°

9 *OABC* is a rectangle. $\overrightarrow{OA} = a$ and $\overrightarrow{OC} = c$. If $|a| = 12$ and $|c| = 5$, then $\angle OMA$, where M is the mid-point of OB, equals (to the nearest degree):

   A 135°   **B** 134°   **C** 136°   **D** 45°

10 If *PQRS* is a cyclic quadrilateral, then $2\overrightarrow{PQ} + \overrightarrow{QR} + \overrightarrow{RS} + \overrightarrow{SP}$ simplifies to:

   A 0   **B** $\overrightarrow{PS}$ **C** $\overrightarrow{QP}$ **D** $\overrightarrow{PQ}$

11 **r** and **s** are two perpendicular vectors, such that $|r| = 5$ and $|s| = 12$. Find:
   (a) $|2r| + |2s|$ (b) $2|r + s|$

12 **x** and **y** are non-parallel vectors and $4x + (h - 3)y = kx + 4y$; find h and k.

13 *PQRST* is a plane five-sided figure. Express as a single vector

$$\overrightarrow{PQ} + \overrightarrow{QR} + \overrightarrow{RS} + \overrightarrow{ST} + 4\overrightarrow{PT}$$

14 If $\mathbf{a} = \begin{pmatrix} 3 \\ y \end{pmatrix}$, $\mathbf{b} = \begin{pmatrix} 2x \\ -4 \end{pmatrix}$ and $2\mathbf{a} + 3\mathbf{b} = \begin{pmatrix} y \\ x \end{pmatrix}$, find y and x.

15 In the parallelogram $WXYZ$ shown in Fig. 7.8, M and N are the mid-points of WX and ZY respectively. P and Q divide WZ and XY respectively in the ratio $1 : 2$. WM represents the vector $\mathbf{u}$ and WP represents the vector $\mathbf{v}$.

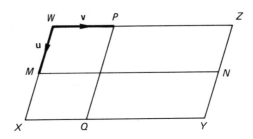

Fig. 7.8

(a) Express in terms of $\mathbf{u}$ and $\mathbf{v}$ the vectors represented by (i) $\overrightarrow{PZ}$ (ii) $\overrightarrow{WY}$ (iii) $\overrightarrow{YP}$

(b) Using only the letters in the diagram, write down the directed line segment which represents the vector $2\mathbf{v} - \mathbf{u}$. [SEB]

16 Given that $\overrightarrow{OP} = \mathbf{p}$, $\overrightarrow{OQ} = \mathbf{q}$ and M is the mid-point of PQ,

(a) express, in terms of $\mathbf{p}$ and $\mathbf{q}$, the vectors $\overrightarrow{PQ}$, $\overrightarrow{PM}$ and $\overrightarrow{OM}$.

Given also that G is the point on OM such that $OG = \frac{2}{3}OM$,

(b) express $\overrightarrow{OG}$ and $\overrightarrow{PG}$ in terms of $\mathbf{p}$ and $\mathbf{q}$.

Given, further, that N is the mid-point of OQ,

(c) express $\overrightarrow{PN}$ in terms of $\mathbf{p}$ and $\mathbf{q}$.

(d) Hence show that $\overrightarrow{PG} = k\overrightarrow{PN}$, where k is a constant, and find the numerical value of k.

(e) State the geometrical meaning of your answer to (d). [L]

17 The position of a ship with respect to a fixed origin O is given by

$$\mathbf{r} = \begin{pmatrix} 0 \\ 10 \end{pmatrix} + t \begin{pmatrix} 0.4 \\ -0.3 \end{pmatrix},$$ where t is the time in minutes after noon, and distances are measured in kilometres.

(a) Calculate the position of the ship when $t = 0, 5, 10, 15, 20$.

(b) Using a scale of 2 cm to 1 km on both the x- and y-axes, draw a diagram to show the path of the ship during the period between noon and 12.20 p.m.

(c) A torpedo has a position given by

$$\mathbf{r} = \begin{pmatrix} 2 \\ 0 \end{pmatrix} + t \begin{pmatrix} 0.5 \\ 0.8 \end{pmatrix}.$$

Using the same set of axes as in part (b), draw a diagram to show the path of the torpedo.

(d) Measure the distance between the ship and the torpedo at 12.05 p.m.

(e) At what time does the ship cross the path of the torpedo? [OLE]

18 $OABCDE$ is a regular hexagon. Let $\overrightarrow{OA} = \mathbf{a}$ and $\overrightarrow{OB} = \mathbf{b}$. Write the following vectors in terms of $\mathbf{a}$ and $\mathbf{b}$:

(a) $\overrightarrow{AB}$; (b) $\overrightarrow{OC}$; (c) $\overrightarrow{BC}$; (d) $\overrightarrow{OD}$; (e) $\overrightarrow{OE}$; (f) $\overrightarrow{AE}$.

Given that $h\mathbf{b} = \mathbf{a} + k(\mathbf{b} - 3\mathbf{a})$, where h and k are numbers, find h and k.

AE meets OB at X. Calculate $OX : OB$. [OLE]

19 In the triangle OAB (Fig. 7.9), $\overrightarrow{OA} = \mathbf{a}$ and $\overrightarrow{OB} = \mathbf{b}$. The points P on OA, Q on AB and R on OB produced are such that $OP = \frac{3}{4}OA$, $AQ = \frac{7}{12}AB$ and $OR = \frac{21}{16}OB$.

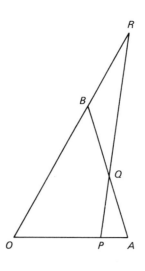

Fig. 7.9

(a) Express   (i) $\overrightarrow{OP}$ in terms of **a**;
                 (ii) $\overrightarrow{OR}$ in terms of **b**;
                (iii) $\overrightarrow{AB}$ in terms of **a** and **b**;
                (iv) $\overrightarrow{AQ}$ in terms of **a** and **b**.
(b) By considering the triangle PAQ, show that $\overrightarrow{PQ} = \frac{7}{12}\mathbf{b} - k\mathbf{a}$, and state the value of k.
(c) By considering the triangle POR, show that $\overrightarrow{PR} = m\mathbf{b} - \frac{3}{4}\mathbf{a}$, and state the value of m.
(d) Find the value of $\dfrac{PQ}{PR}$.   [EAUC]

20 In Fig. 7.10, $\overrightarrow{OA} = \mathbf{a}$, $\overrightarrow{OB} = \mathbf{b}$, $\overrightarrow{OC} = \mathbf{c}$ and $\overrightarrow{OD} = \mathbf{d}$. X is the mid-point of AB and Y is the mid-point of CD.

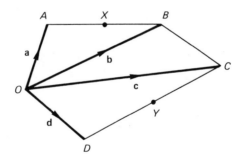

Fig. 7.10

(a) Find, in terms of **a** and **b**: (i) $\overrightarrow{AB}$; (ii) $\overrightarrow{AX}$; (iii) $\overrightarrow{OX}$.
(b) Similarly, find $\overrightarrow{OY}$ in terms of **c** and **d**.
(c) If M is the mid-point of XY (not shown in the figure), write down an expression for $\overrightarrow{OM}$ in terms of **a**, **b**, **c**, and **d**.
(d) Describe exactly the position of P if $\overrightarrow{OP} = \frac{1}{2}(\mathbf{a} + \mathbf{d})$.   [EAUC]

21 In Fig. 7.11, E is the point on the side BD of the triangle ABD such that $DE = kDB$. The lines AB and DC are parallel and $DC = mAB$.
    It is given that $\overrightarrow{AB} = 10\mathbf{a}$ and $\overrightarrow{AD} = 2\mathbf{b}$.
(a) Express $\overrightarrow{DB}$ in terms of **a** and **b**.
(b) Express $\overrightarrow{AE}$ in terms of **a**, **b** and k.

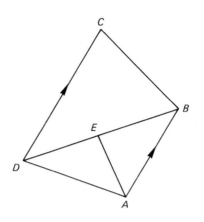

Fig. 7.11

(c) Express $\overrightarrow{BC}$ in terms of **a**, **b** and m.

(d) If BC is parallel to AE, show that $m - 1 - km = 0$.

22 $ABCDEF$ is a regular hexagon. The displacements $\overrightarrow{AB}$ and $\overrightarrow{BC}$ are denoted by **x** and **y** respectively.

  (a) Express each of the following displacements in terms of **x** or **y** or both:
  (i) $\overrightarrow{DE}$ (ii) $\overrightarrow{FC}$ (iii) $\overrightarrow{AF}$ (iv) $\overrightarrow{FD}$ (v) $\overrightarrow{EC}$

  (b) If EC and AB are produced to meet at G, explain why $BFCG$ is a parallelogram.

  Hence, or otherwise, express in terms of **x** or **y** or both:
  (i) $\overrightarrow{AG}$ (ii) $\overrightarrow{EG}$

  (c) The area of the hexagon is 6 units. Find the areas of:
  (i) triangle BCG (ii) quadrilateral $AGEF$ [NI]

23 The square $OACB$ (Fig. 7.12) is divided into 16 equal squares of side 1 cm. Given that $\overrightarrow{OA} = 4\mathbf{a}$ and $\overrightarrow{OB} = 4\mathbf{b}$, draw a diagram and mark clearly

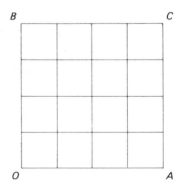

Fig. 7.12

  (a) the point X such that $\overrightarrow{OX} = \mathbf{a} + 3\mathbf{b}$;

  (b) the point Y such that $\overrightarrow{YA} = 2\mathbf{a} - \mathbf{b}$;

  (c) a point Z, at one of the intersections, such that $|\overrightarrow{OZ}|$ 5 cm.

24 (a) The vectors **e** and **n** represent velocities of 1 km/h due east and 1 km/h due north respectively. A man rows a boat on a river in such a way that, if there were no current, its velocity would be $2\mathbf{n} - \mathbf{e}$. On a river flowing with velocity $3\mathbf{e}$ the boat's actual velocity is **v**.

  (i) Express **v** in the form $a\mathbf{e} + b\mathbf{n}$.

  (ii) Calculate $|\mathbf{v}|$.

122

(b) In the triangle ABR (Fig. 7.13), T is the mid-point of AB and S is the mid-point of BR. Given that $\mathbf{AS} = \mathbf{s}$ and $\mathbf{AT} = \mathbf{t}$, express in terms of $\mathbf{s}$ and/or $\mathbf{t}$, as simply as possible, the vectors

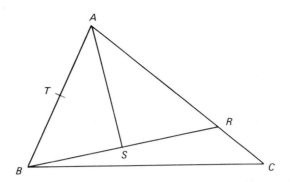

Fig. 7.13

(i) $\mathbf{TS}$ (ii) $\mathbf{AB}$ (iii) $\mathbf{BR}$ (iv) $\mathbf{AR}$.

Given also that C is the point such that $\mathbf{AR} = 5\mathbf{RC}$ and $\mathbf{TS} = k\mathbf{RC}$, find the value of k. [UCLES]

25 (a) Given that $\mathbf{OK} = \begin{pmatrix} 16 \\ 2 \end{pmatrix}$, $\mathbf{OL} = \begin{pmatrix} 4 \\ -3 \end{pmatrix}$ and that M and N are the mid-points of OK and OL respectively,

(i) express $\mathbf{MN}$ as a column vector;
(ii) find the value of $|\mathbf{KL}|$.

(b) Referring to Fig. 7.14, $\mathbf{OA} = 4\mathbf{a}$, $\mathbf{OB} = 4\mathbf{b}$ and $\mathbf{BP} = 3\mathbf{a} - \mathbf{b}$. Express as simply as possible, in terms of $\mathbf{a}$ and $\mathbf{b}$:

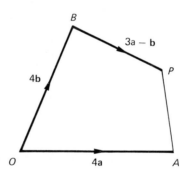

Fig. 7.14

(i) $\mathbf{OP}$ (ii) $\mathbf{AP}$.

The lines OA produced and BP produced meet at Q. Given that $\mathbf{BQ} = m\mathbf{BP}$ and $\mathbf{OQ} = n\mathbf{OA}$, form an equation connecting m, n, $\mathbf{a}$ and $\mathbf{b}$. Hence deduce the values of m and n. [UCLES]

8 Transformation Geometry

8.1   Translation

In Fig. 8.1, each point of *P* has been mapped onto *Q*, by a simple *displacement* or *translation*. A translation can be represented by a vector.

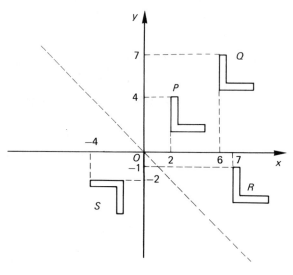

Fig. 8.1

We write $P \rightarrow Q$ under the translation $\mathbf{T}_1 = \begin{pmatrix} 4 \\ 3 \end{pmatrix}$. Similarly, $Q \rightarrow R$ under the translation $\mathbf{T}_2 = \begin{pmatrix} 1 \\ -8 \end{pmatrix}$.

If we wish to carry out one transformation followed by another, i.e. $\mathbf{T}_1$ followed by $\mathbf{T}_2$, it is denoted by $\mathbf{T}_2 \mathbf{T}_1$. Note the order of letters.

Now $P \rightarrow R$ under the translation $\mathbf{T}_3 = \begin{pmatrix} 5 \\ -5 \end{pmatrix}$ but $\mathbf{T}_3$ is the same as $\mathbf{T}_2 \mathbf{T}_1$. It can be seen that combining translations is the same as adding vectors.

$$\therefore \ \mathbf{T}_2 \mathbf{T}_1 = \begin{pmatrix} 1 \\ -8 \end{pmatrix} + \begin{pmatrix} 4 \\ 3 \end{pmatrix} = \begin{pmatrix} 5 \\ -5 \end{pmatrix} = \mathbf{T}_3 .$$

8.2   Reflection

Referring again to Fig. 8.1, $P \rightarrow S$ by a *reflection* in the line $y = -x$. This line is called the *mirror line*. Reflection in a line through the origin can be represented by using a 2 x 2 matrix; see section 9.13.

8.3 Rotation

In Fig. 8.2, the flag motif $X \to Z$ by a rotation $\mathbf{R}_1$, centre $(1, 1)$, angle $-90°$. (**Note that a clockwise rotation is negative.**)

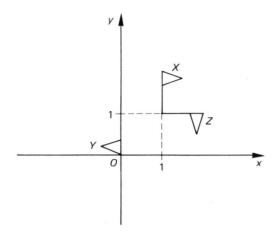

Fig. 8.2

Also, $X \to Y$ by a rotation $\mathbf{R}_3$, centre $(\frac{1}{2}, 1)$, angle $180°$, and $Z \to Y$ by a rotation $\mathbf{R}_2$, centre $(\frac{1}{2}, 1\frac{1}{2})$, angle $-90°$. (Check this by accurate drawing.)

$$\therefore \mathbf{R}_3 = \mathbf{R}_2 \mathbf{R}_1.$$

In most cases, a *rotation* followed by a *rotation* is a *rotation*.
Try to find an example of when this is not true.

A rotation about O can be represented by a 2 x 2 matrix; see section 9.13.

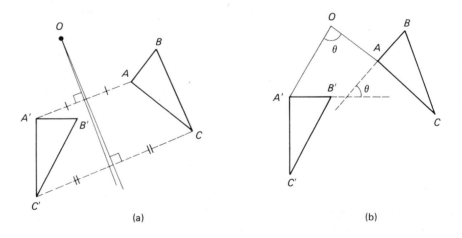

(a) (b)

Fig. 8.3

To find the centre of rotation by construction (see Fig. 8.3(a)):

Join two pairs of corresponding points AA' and CC'. Bisect each line at right angles, and where these bisectors meet is the centre of rotation O. Note also in Fig. 8.3(b) all lines in the shape are rotated by the angle of rotation.

A reflection followed by a reflection can often be represented by a rotation.

In Fig. 8.4 a reflection in m_1 followed by a reflection in m_2 is equal to a rotation of $2x°$ twice the angle between the mirror lines about O (the point of intersection of the mirror lines).

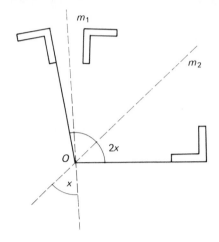

Fig. 8.4

8.4   Glide Reflection

A *glide reflection* is the name given to a simple transformation which consists of a reflection followed by a translation, or vice versa. See Fig. 8.5.

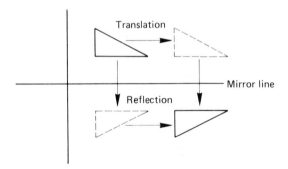

Fig. 8.5

8.5   Isometries

The transformations: reflection, rotation, translation and glide reflection all leave the shape of the transformed figure unchanged. For this reason, they are called *isometries*.

8.6   Dilatation

Figure 8.6 shows two examples of a dilatation, one an enlargement, the other a reduction.

LEE DIXON
ARSENAL

LITTLEWOODS CUP
FINAL SPECIAL!
IN COLOUR
ARSENAL AND LUTON TEAM GROUPS!

Emlyn Hughes' exclusive verdict!

PLUS

'Armchair Guide' and 'How They Got There'!

'Private Life' double – Michael Thomas and Danny Wilson.

Exclusive 'Match Facts' form guide!
INTERVIEWS WITH THE STARS!
BUMPER CUP FINAL QUIZ

NEXT WEEK

IN THE FOOTBALL MAGAZINE WHERE THE ACTION NEVER STOPS....

GREAT NEW SERIES!
EUROPEAN CHAMPIONSHIP COUNTDOWN

No.1
DENMARK

THE MARK LAWRENSON INTERVIEW

PLUS

International previews featuring Hungary v England and Eire v Yugoslavia

MINSTER MEN MISERY!
The football first no-one wanted.

Editor: Melvyn Bagnall
News Editor: Paul Stratton
Reporters: Adrian Curtis, Howard Wheatcroft, Ray Ryan
Production Editor: Mick Weavers
Design: John Salter
Photographer: Philip Bagnall
Editor's Secretary: Jacquie Apthorpe
Advertisement Manager: Brian Reacher
Advertisement Assistant: Mike Wells
Product Manager: Denis Stapleton
Publisher: Ken Gill
Premium Sales Manager: Neil Pitcher

Vol. 9 No. 31

MATCH

Editorial: Stirling House, Bretton, Peterborough PE3 8DJ. Tel: 0733 260333/264666
Advertising and Marketing: Bretton Court, Bretton, Peterborough PE3 8DZ. Tel: 0733 264666

Reader Offers: P.O. Box 136, Peterborough PE2 0XW. Tel: 0733 237111
Circulation and Back Issues: EMAP Frontline, 1 Lincoln Court, Lincoln Road, Peterborough PE1 2RP. Tel: 0733 555161
Subscriptions: P.O. Box 500, Leicester LE99 0AA. Orderline 0858 410888
Mono origination: Typefont Ltd
Colour origination: Lumarcolour Ltd
Printing: Chase Web, Barnstaple
© EMAP Pursuit Publishing Ltd 1988
Registered at the Post Office as a newspaper

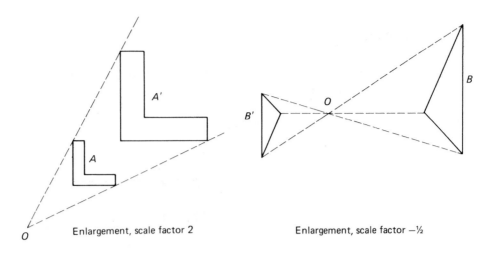

Fig. 8.6

All corresponding lengths in Fig. 8.6(a) are multiplied by 2, even distances from the centre of enlargement O. All corresponding lengths in Fig. 8.6(b) are multiplied by $\frac{1}{2}$. The negative sign indicates that the centre of enlargement is between the object and the image.

8.7   Shear

Students often find it difficult to grasp the concept of a shear. In Fig. 8.7(a), $A \to A'$ by a shear parallel to the x-axis, and the x-axis is *invariant*. The further a point is from the invariant line, the greater the shift of the point. In order to define a shear, specify where one point moves, and the invariant line.

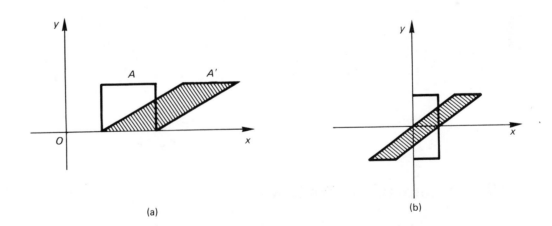

Fig. 8.7

The distance moved by any point in the direction of the invariant line is proportional to the distance from the invariant line. The constant of proportionality is called the *shear factor*.

In Fig. 8.7(b) the same shear has been applied to a shape part of which lies *below* the invariant line. It can be seen that the part below moves backwards, because the distance from the invariant line is negative. A shear can be represented by a 2 x 2 matrix. Shearing preserves the area of a shape.

127

8.8   Stretch

In Fig. 8.8, the square $OABC$ has been stretched by a factor 3 in a direction parallel to the x-axis.

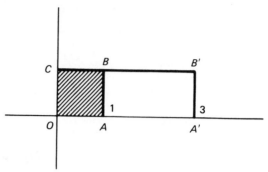

Fig. 8.8

8.9   Similarity and Congruence

If two shapes are *exactly* the same but have a different orientation, they are said to be *congruent*.

If one shape is a simple enlargement of another, in two or three dimensions, they are said to be *similar*.

Situations where similar shapes occur are shown in Fig. 8.9.

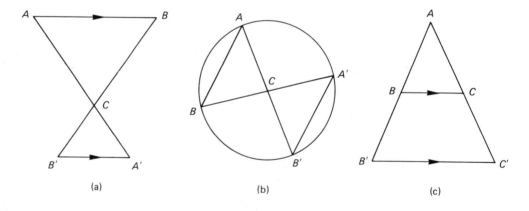

(a)   (b)   (c)

Fig. 8.9

8.10   Geometrical Problems

Modern geometry use the ideas of transformation geometry to prove geometrical theorems, and to carry out calculations. See section 12.2 for a proof of Pythagoras' theorem. We illustrate these ideas with the following examples; a full treatment is beyond the scope of this book.

Worked Example 8.1

The points A and B have coordinates (2, 3) and (4, 4) respectively. Their images under a rotation are A' (3, −4) and B' (4, −6). Plot A and B and their images on graph paper and by construction or calculation determine the centre and angle of the rotation.

It is also possible to transform A and B to A' and B' by reflection in $y = x$, followed by reflection in a second line. Find the equation of the second line.

[OLE]

Solution

The centre of rotation lies on the line $y = -1$. This is the perpendicular bisector of BB' see Fig. 8.10. The perpendicular bisector of AA' meets this at the point $(-1, -1)$.

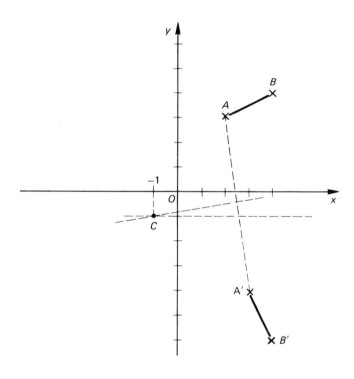

Fig. 8.10

The centre of rotation is $C(-1, -1)$.

The angle of rotation is the angle between AC and $A'C$ by measurement $= 90°$.

If B is reflected in the line $y = x$ it maps onto $(4, 4)$ (i.e. it stays put).

$$\therefore \text{ Since } (4, 4) \rightarrow (4, -6).$$

The other line of reflection must be $y = -1$.

Worked Example 8.2

Using a scale of 1 cm to 1 unit on both axes, plot the points $P(-5, -1)$, $Q(-3, -2)$, $R(6, 4)$ and $S(5, 1)$. Draw the triangles OPQ and ORS.

A transformation, T, consists of a reflection in a line through O followed by an enlargement, centre O, such that triangle OPQ is transformed by T onto triangle ORS.

(a) Use ruler and compasses only to construct a suitable axis of reflection.

(b) The equation of that axis of reflection may be written in the form $y = kx$.

Estimate from your graph the value of k to two significant figures.

(c) Calculate the lengths of *OP* and *OR* and hence deduce the linear scale factor of the enlargement.

(d) Write down the coordinates of the images of points *P* and *Q* when triangle *OPQ* is transformed by T^2.   [OLE]

Solution

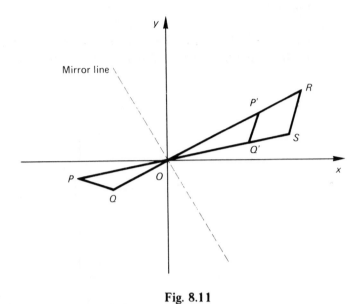

Fig. 8.11

(a) Mark off length *OP* on *OR* and length *OQ* on *OS*. This gives the images *P′* and *Q′* after reflection.

(b) The mirror line at right angles to *QQ′*, can be drawn in. Its gradient is −2.4

$$\therefore k = -2.4.$$

$$OP = \sqrt{5^2 + 1^2} = \sqrt{26}, \, OR = \sqrt{6^2 + 4^2} = \sqrt{52}.$$

The scale factor of the enlargement is $\dfrac{OR}{OP} = \dfrac{\sqrt{52}}{\sqrt{26}} = \sqrt{2}.$

After T^2, the enlargement is by a factor 2, and the triangle has been reflected back to the third quadrant.

$$\therefore \text{Image of } P \text{ is } (-10, -2),$$
$$\text{image of } Q \text{ is } (-6, -4).$$

Worked Example 8.3

In Fig. 8.12 *M* and *N* are the mid-points of the sides *AB* and *AC* respectively of

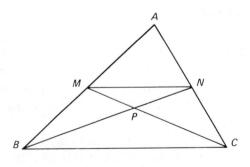

Fig. 8.12

triangle *ABC*, which is not isosceles. *BN* and *CM* meet at *P*. An enlargement has centre *A* and scale factor 2.

(a) Under this enlargement, which point is the image of (i) *M*, (ii) *N*?

(b) Let the images of *B*, *C* and *P* under this enlargement be *B'*, *C'* and *P'* respectively. Draw a neat copy of the diagram and mark the positions of *B'*, *C'* and *P'*.

(c) Which triangle in your figure is similar to triangle *ABC'*?

(d) Describe the shape of the quadrilateral *BPCP'*.

(e) *L* is the point at which *AP* produced meets *BC*. Give a reason why *L* is the mid-point of *BC*.

(f) Find the ratios of *AP* : *PL* and *BC* : *MN*, giving reasons for your answers.

(g) Complete the following:

$$\frac{BC}{MN} = \frac{}{MP} = BP$$

(h) State the ratio of *BP* : *PN*. [OLE]

Solution

(a) (i) *B*; (ii) *C* since *AB* = 2*AM* and *AC* = 2*AN*.

(b) See Fig. 8.13.

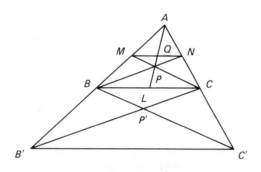

Fig. 8.13

(c) △*AMC* is similar to *ABC'*, because *AB* = 2*AM*, *AC'* = 2*AC* and angle *BAC* is common to both triangles.

(d) *MC* // *BC'* and *BN* // *B'C*
Hence *BPCP'* is a parallelogram.

(e) This is a little awkward at first sight. The reason is that the medians of a triangle (lines joining vertices to the mid-points of the opposite sides) all meet at a point. Since *CM* and *BN* are medians, *AP* must be a median, therefore *L* is the mid-point.

(f) If *AL* meets *MN* at *Q*, and if we let *AQ* = *x*, then *QL* = *x*.
△*BPC* is an enlargement of △*NPM*, scale factor 2, therefore

$$PL = \frac{2x}{3}, \quad QP = \frac{1}{3}x$$

$$\therefore AP = AQ + QP = x + \frac{1}{3}x = \frac{4x}{3}.$$

Hence $AP : PL = \frac{4x}{3} : \frac{2x}{3} = 2 : 1$

131

$BC : MN = 2 : 1$ enlargement, centre A, scale factor 2.

(g) $\dfrac{BC}{MN} = \dfrac{BP'}{MP} = \dfrac{BP}{PN}$. All ratios are equal to 2.

(h) $BP : PN = 2 : 1$.

Worked Example 8.4

In Fig. 8.14, OL and OM are two lines inclined to one another at an angle $\beta°$, and P is a point such that angle $POL = \alpha°$. P' is the image of P on reflection in OL, and P'' is the image of P' on reflection in OM. The angle POP'', in degrees, is

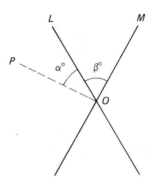

Fig. 8.14

A $\alpha + \frac{3}{2}\beta$ **B** $\beta - \alpha$ **C** $\alpha + 2\beta$ **D** 2β **E** $2\alpha + \beta$ [OLE]

Solution

If α is less than β, then P' lies between OL and OM. See Fig. 8.15(a).
The angle $POP'' = \alpha + \alpha + \beta - \alpha + \beta - \alpha = 2\beta$.
If α is greater than β, then P' lies outside OL and OM. See Fig. 8.15(b).
The angle $POP'' = \alpha + \beta - (\alpha - \beta) = \alpha + \beta - \alpha + \beta = 2\beta$.

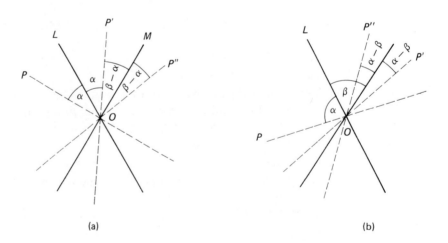

(a)   (b)

Fig. 8.15

In each case the correct answer is **D**.

8.11 Area and Volume of Similar Figures

If a shape is enlarged by a scale factor k, its area is enlarged by a scale factor k^2.

If a three-dimensional shape is enlarged by a scale factor k, its volume is enlarged by a scale factor of k^3.

Note also that its surface area is enlarged by a scale factor of k^2.

The next three examples illustrate quite popular examination questions.

Worked Example 8.5

Figure 8.16 shows a parallelogram $PQRS$, and T is the mid-point of PQ. Find the following ratios:

(a) $\dfrac{\text{Area of triangle } TQU}{\text{Area of triangle } SUR}$;

(b) $\dfrac{\text{Area of triangle } TQU}{\text{Area of quadrilateral } PTUS}$.

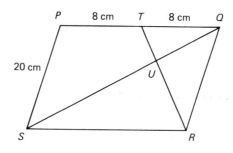

Fig. 8.16

Solution

Since no angles are given, it is not possible to work out any of the individual areas. The methods of similar figures must be used.

(a) $\triangle SUR$ is an enlargement of $\triangle QUT$ centre U and scale factor -2 ($SR = 16$ cm, $TQ = 8$ cm);

hence $\qquad\qquad \dfrac{\text{Area } \triangle TQU}{\text{Area } \triangle SUR} = \dfrac{1}{(-2)^2} = \tfrac{1}{4}$.

(b) There is no direct enlargement from $\triangle TQU$ to the quadrilateral $PTUS$, hence a different technique must be used. Although the following method is not the only one, it has the advantage of being easily seen diagrammatically. In Fig. 8.17 the area of $\triangle TQU$ is denoted by A. Hence the area of $\triangle SUR$ is $4A$.

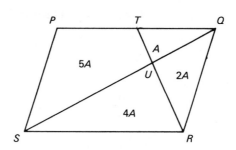

Fig. 8.17

133

$UR = 2TU$, hence if you regard TU as the base of $\triangle TUQ$ and UR as the base of $\triangle URQ$, they have the same height, but the base has been doubled. Hence the area has been doubled. The area of $\triangle UQR$ is therefore $2A$. Since the diagonal SQ divides the area of the area of the parallelogram into two equal parts, the area of quadrilateral $PTUS$ must be $5A$.

Hence
$$\frac{\text{Area } \triangle TQU}{\text{Area } PTUS} = \frac{A}{5A} = \frac{1}{5}.$$

Worked Example 8.6

A cylinder has a volume of 100 cm^3. If another cylinder of the same material is made with half the radius, and three times the height of the original cylinder, find the volume of the second cylinder.

Solution

In this problem, it would appear that there are two unknowns, the height and radius of the original cylinder. Also, we are not given a value for π.
Let R cm be the radius of the first cylinder.
Let H cm be the height of the first cylinder.
The volume of a cylinder is given by the formula $V = \pi r^2 h$,

$$\therefore \ 100 = \pi R^2 H. \tag{1}$$

For the new cylinder, $r = \frac{1}{2}R$ and $h = 3H$.

The new volume
$$V = \pi (\tfrac{1}{2}R)^2 (3H) = \frac{3\pi R^2 H}{4}. \tag{2}$$

Using equation (1), however,

$$V = \tfrac{3}{4} \times 100 = 75.$$

The new volume is 75 cm^3.

Worked Example 8.7

A balloon is filled with hydrogen, so that its volume is V cm^3. The radius of the balloon is then 6 cm. The balloon is then further inflated so that its volume increases by 40 per cent. Find the new radius of the balloon. Assume the shape of the balloon is a sphere at all times.
    The method of enlargements is always easier to use in this type of problem. Find the scale factor for the quantities that you know.

Solution

Original volume $= V$ cm^3.
New volume $= V + \dfrac{40V}{100} = 1.4V$ cm^3.

$$\frac{\text{New volume}}{\text{Old volume}} = \frac{1.4V}{V} = 1.4.$$

But $\left(\dfrac{\text{New radius}}{\text{Old radius}}\right)^3 = \dfrac{\text{New volume}}{\text{Old volume}} = 1.4.$

Take the cube root of each side:

$$\dfrac{\text{New radius}}{\text{Old radius}} = \sqrt[3]{1.4} = 1.12.$$

$$\text{New radius} = 6 \times 1.12$$

$$= 6.72 \text{ cm}.$$

Exercise 8

1 Given that two triangles are similar, it then follows that:
   I the areas of the triangles are in the same ratio as the squares of corresponding sides;
   II the three sides of one triangle are equal in length to the lengths of the corresponding sides of the other;
   III the three interior angles of one triangle are equal to the corresponding angles in the other.

   A   I and II only   **B**   I and III only
   C   II and III only   **D**   I, II and III   [AEB]

2 In Fig. 8.18, *AE* equals:

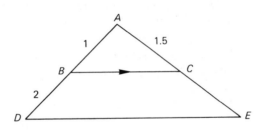

Fig. 8.18

   A   3   **B**   3.5   **C**   4   **D**   4.5

3 *ABC* is a triangle and *E* is a point on *AB* such that *AE* : *EB* = 2 : 5. *F* is on *AC* and *EF* is parallel to *BC*. It follows that the ratio of the area of trapezium *BEFC* to the area of the triangle *AEF* is:
   A   4 : 25   **B**   21 : 25   **C**   45 : 4   **D**   2 : 5

4 A metal sphere of diameter 150 mm has a mass of 120 g. A metal sphere of the same material, which weighs 60 g, will have a diameter of approximately:
   A   119 mm   **B**   75 mm   **C**   37.5 mm   **D**   18.8 mm

5 Sphere *X* has diameter 5 cm and sphere *Y* has radius 6 cm. The ratio of their surface areas is:
   A   25 : 36   **B**   36 : 25   **C**   25 : 24   **D**   25 : 144

6 A translation followed by a reflection will always be:
   A   a reflection   **B**   a rotation
   C   a translation   **D**   none of these

7 The coordinates of the point *P* (3, 1) after reflection in the line $y + x = 0$, followed by reflection in the line $y = 0$, are:
   A   (3, −1)   **B**   (−1, 3)   **C**   (1, −3)   **D**   (−1, −3)

8 In Fig. 8.19, *BOC* and *AOD* are straight lines, $2OC = OB$ and $2OD = AO$, *E* is the mid-point of *OC* and *F* the mid-point of *OB*. Describe the transformation

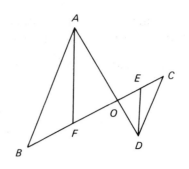

Fig. 8.19

which maps AF into DE. What is the ratio of the area of $\triangle OED$ to that of $\triangle OAB$?

9 A square $ABCD$ has sides of length 2 cm. M_1 denotes a reflection in AB, M_2 a reflection in BC, M_3 a reflection in CD and M_4 a reflection in DA. A small equilateral triangle of side about $\frac{1}{2}$ cm is placed with one side parallel to AB and about $\frac{1}{2}$ cm from AB outside the square with the opposite vertex of the triangle pointing away from the square. Draw the successive positions of this triangle after the reflections M_1, M_2, M_3 and M_4. Show, by drawing, that the transformation $M_4 M_3 M_2 M_1$ can be replaced by a simple rotation.

10 M stands for a reflection in the line $y = x$ and T stands for a translation of 2 units in the direction of the positive x-axis. Draw diagrams to illustrate the effect of the following transformations on the triangle whose vertices are the points A $(1, 0)$, B $(0, 0)$ and C $(1, 1)$:
   (a) MT;   (b) $T^{-1}MT$;   (c) MTM.

11 In Fig. 8.20, $ABCD$ is a rectangle, $AX = DC$. DY is parallel to XB. Prove that triangles ANX and CPD are congruent.

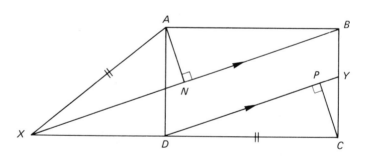

Fig. 8.20

Hence prove that NP is equal and parallel to XD.

12 M_1 is a reflection in the line $y = 3$ and M_2 is a reflection in the line $x + y = 4$. By considering the image of the point A $(2, 3)$ under the transformations (i) $M_1 M_2$, (ii) $M_2 M_1$, show that $M_1 M_2 \neq M_2 M_1$.

Show that $M_1 M_2$ and $M_2 M_1$ can each be replaced by a single rotation with the same centre of rotation. What is this centre of rotation? If these two rotations are denoted by R_1 and R_2 respectively, what is the image of A under the transformation $R_1 R_2$?

13 The following list defines a number of transformations:

R is a rotation of 90° anticlockwise about the origin.
H is a rotation of 180° about the origin.
M_x is a reflection in the x-axis.

M_y is a reflection in the y-axis.
T_x is a translation of 2 units parallel to the x-axis.
T_y is a translation of 2 units parallel to the y-axis.

The ends A and B of a line segment have coordinates (2, 0) and (2, 4) respectively. Show, by clear diagrams, that the effect of the transformation $T_x M_y T_y R$ is to move AB to a position BC. State the coordinates of C.

Show also that $RT_y M_y T_x$ produces a different result, but that $KRT_y M_y T_x$ will transform AB to BC if K is one of the transformations in the list above. Which transformation is K?

14 A, B and C are the points (2, 4), (−3, −6) and (5, −2) respectively.
   (a) Write down the components of the translation BC and calculate its magnitude.
   (b) The translation which maps B onto C maps A onto D. State the coordinates of D.
   (c) Under a dilatation centre the origin B is the image of A. State the scale factor of the dilatation.   [SEB]

15 Figure 8.21 shows two unequal squares, $OABC$ and $OPQR$, with centres H and K respectively. Draw a diagram showing the squares and the image of triangle COR when it is rotated through 90° clockwise about H.

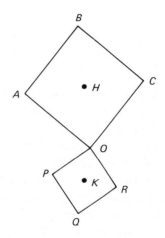

Fig. 8.21

Let T be the image of R under the same rotation. Prove that angle $TOP =$ angle CRO.

State the single transformation which maps triangle COR onto triangle TPO.

Let AP meet OT at X. Prove that X is the mid-point of AP. Prove that $2OX = CR$.   [OLE]

16 A triangle ABC, whose vertices are A (−1, 2), B (−2, 3) and C (−4, 2), is subjected to a geometrical transformation R followed by another transformation T, such that A, B and C are mapped onto A' (4, −1), B' (5, 0) and C' (4, 2) respectively. It is given that R is a rotation about O (0, 0) and T is a translation.
   (a) On graph paper, draw the triangles ABC and $A'B'C'$. On the same diagram, draw the image of triangle ABC under the transformation R.
   (b) State the angle of rotation of the transformation R and find the matrix associated with this transformation.
   (c) State the vector of the translation T.
   (d) Describe geometrically the single transformation which would map triangle ABC onto triangle $A'B'C'$.   [JMB]

17 Taking the origin O near the centre of a sheet of $\frac{1}{2}$ cm square paper, plot the points A (2, 0), B (0, 4) and C (−4, 2).

(a) On your diagram, complete the square $ABCD$ and write down the co-ordinates of the fourth vertex D.

(b) Under a half-turn about the point (3, −3), $A \to P$, $B \to Q$, $C \to R$ and $D \to S$. Plot the points P, Q, R and S on your diagram and write down their coordinates.

(c) Under the dilatation $[O, 2]$, $A \to K$, $B \to L$, $C \to M$ and $D \to N$. Plot the points K, L, M and N on your diagram and write down their coordinates.

(d) State the coordinates of the centre and the scale factor of the dilatation under which $K \to P$, $L \to Q$, $M \to R$ and $N \to S$. [SEB]

18 In Fig. 8.22, $PQRS$ is a trapezium with PQ parallel to SR. $\angle PST = 90°$. $PQ = 6$ cm, $PS = 4$ cm, $TR = 1.5$ cm and $QU = 3$ cm.

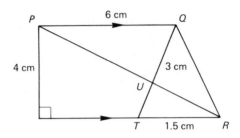

Fig. 8.22

(a) What can be said about triangles PUQ and TUR?

(b) What can be said about the areas of triangles TPR and TQR?

(c) Calculate TU.

(d) Find in its simplest form the ratio of areas of $\triangle TUR : \triangle PUQ$.

(e) Find, in its simplest form, the ratio of area of $\triangle PUT : \triangle QUP$.

(f) Find the area of the trapezium $PQRT$.

19 On graph paper, using a scale of 1 cm to represent 1 unit on each axis, draw the square whose vertices are O (0, 0), A (1, 0), B (1, 1) and C (0, 1). Calculate the coordinates of the images O', A', B' and C' of these points under the transformation M whose matrix is M, where

$$M = \begin{pmatrix} 4 & -3 \\ 3 & 4 \end{pmatrix}.$$

Draw the image figure $O'A'B'C'$ on your graph.

Using Pythagoras' theorem or otherwise, calculate $O'A'$ and $O'C'$. Measure $\angle A'O'C'$. Hence calculate the area $O'A'B'C'$, and state a relationship between the ratio $\dfrac{\text{area } O'A'B'C'}{\text{area } OABC}$ and the matrix M.

Find the coordinates of the points P and Q where the line $3y - 4x = 6$ cuts the x-axis and the y-axis respectively. Calculate the images P and Q of these points under the transformation M. Find the tangent of the angle between PQ and $P'Q'$. [L]

9 Matrices

9.1 Array Storage

A great deal of information in everyday life can be represented in a two-dimensional table, called a *matrix*. Table 9.1 shows the milkman's weekly order from one household. The information is compact, and easy to read. This type of *information storage* is commonly used in computers.

Table 9.1

	Mon	Tue	Wed	Thur	Fri	Sat	Sun
Pints of milk	2	2	3	2	1	1	3
Cartons of cream	0	0	1	0	0	0	1
Eggs	6	0	0	0	6	0	0

In this example, the matrix has three *rows* and seven *columns*. We say that the *order* of the matrix is 3×7.

The matrix can be written

$$\begin{pmatrix} 2 & 2 & 3 & 2 & 1 & 1 & 3 \\ 0 & 0 & 1 & 0 & 0 & 0 & 1 \\ 6 & 0 & 0 & 0 & 6 & 0 & 0 \end{pmatrix}$$

if we leave out the headings. The numbers in a matrix are called its *elements*.

9.2 Sum Difference, Multiplication by a Scalar

If $A = \begin{pmatrix} 3 & 4 \\ 1 & 6 \\ 2 & 5 \end{pmatrix}$ and $B = \begin{pmatrix} 4 & -1 \\ 2 & 0 \\ 3 & 1 \end{pmatrix}$

then $A + B = \begin{pmatrix} 3+4 & 4+-1 \\ 1+2 & 6+0 \\ 2+2 & 5+1 \end{pmatrix} = \begin{pmatrix} 7 & 3 \\ 3 & 6 \\ 5 & 6 \end{pmatrix}$.

To add two matrices, simply add the corresponding element.
Two matrices can only be added if they have the same order.

$B + A$ also $= \begin{pmatrix} 7 & 3 \\ 3 & 6 \\ 5 & 6 \end{pmatrix}$,

$\therefore A + B = B + A$.

Addition of matrices is *commutative*, and also clearly *associative*.

$$A - B = \begin{pmatrix} -1 & 5 \\ -1 & 6 \\ -1 & 4 \end{pmatrix}.$$ Simply subtract the corresponding elements.

$$3A = A + A + A = \begin{pmatrix} 9 & 12 \\ 3 & 18 \\ 6 & 15 \end{pmatrix}.$$ Each element is multiplied by 3.

9.3 Multiplication

To multiply two matrices is a more complicated exercise. The numbers in the following example have been carefully chosen to show the method clearly:

$$\begin{pmatrix} 2 & 5 \\ 6 & 7 \\ 9 & 8 \end{pmatrix} \begin{pmatrix} 10 & 0 & 4 \\ 3 & 16 & 1 \end{pmatrix} = \begin{pmatrix} 2 \times 10 + 5 \times 3 & 2 \times 0 + 5 \times 16 & 2 \times 4 + 5 \times 1 \\ 6 \times 10 + 7 \times 3 & 6 \times 0 + 7 \times 16 & 6 \times 4 + 7 \times 1 \\ 9 \times 10 + 8 \times 3 & 9 \times 0 + 8 \times 16 & 9 \times 4 + 8 \times 1 \end{pmatrix}$$

3 x 2   2 x 3
order   order

$$= \begin{pmatrix} 35 & 80 & 13 \\ 81 & 112 & 31 \\ 114 & 128 & 44 \end{pmatrix}$$

3 x 3
order

Two matrices can only be multiplied if the number of columns in the first matrix equals the number of rows in the second. This means that

$$\begin{pmatrix} 10 & 0 & 4 \\ 3 & 16 & 1 \end{pmatrix} \begin{pmatrix} 2 & 5 \\ 6 & 7 \\ 9 & 8 \end{pmatrix} = \begin{pmatrix} 56 & 82 \\ 111 & 135 \end{pmatrix}.$$

2 x 3   3 x 2   2 x 2

Clearly, multiplication of matrices is not commutative.

See Worked Example 9.2 for an application of matrix multiplication.

9.4 Algebraic Manipulation 1

If $A = \begin{pmatrix} 4 & 6 \\ 3 & 5 \end{pmatrix}$, then $3A = \begin{pmatrix} 12 & 18 \\ 9 & 15 \end{pmatrix}$; each element is multiplied by 3.

If also $B = \begin{pmatrix} 2 & -1 \\ 3 & 5 \end{pmatrix}$,

then $4A - 3B = \begin{pmatrix} 16 & 24 \\ 12 & 20 \end{pmatrix} - \begin{pmatrix} 6 & -3 \\ 9 & 15 \end{pmatrix} = \begin{pmatrix} 10 & 27 \\ 3 & 5 \end{pmatrix}.$

See Check 25.

9.5 Identity

If $I = \begin{pmatrix} 1 & 0 \\ 0 & 1 \end{pmatrix}$ and $A = \begin{pmatrix} 4 & 1 \\ 2 & -4 \end{pmatrix}$, then $IA = \begin{pmatrix} 1 & 0 \\ 0 & 1 \end{pmatrix}\begin{pmatrix} 4 & 1 \\ 2 & -4 \end{pmatrix}$

$$= \begin{pmatrix} 4 & 1 \\ 2 & -4 \end{pmatrix} = AI.$$

The matrix $\begin{pmatrix} 1 & 0 \\ 0 & 1 \end{pmatrix}$ is called the 2×2 *identity* matrix. It leaves a matrix unchanged if multiplied by it.

The matrix
$$\begin{pmatrix} 1 & 0 & 0 & \cdots \\ 0 & 1 & 0 & \cdots \\ 0 & 0 & 1 & \cdots \\ \cdots \cdots \cdots \cdots \end{pmatrix}$$
$$n \times n$$

is the $n \times n$ identity matrix; it has 1's on the *leading diagonal*, and zero everywhere else.

9.6 Inverse Matrices

If $A = \begin{pmatrix} 4 & 3 \\ 5 & 6 \end{pmatrix}$ and $B = \begin{pmatrix} \frac{6}{9} & -\frac{3}{9} \\ -\frac{5}{9} & \frac{4}{9} \end{pmatrix}$,

then $AB = \begin{pmatrix} 4 & 3 \\ 5 & 6 \end{pmatrix}\begin{pmatrix} \frac{6}{9} & -\frac{3}{9} \\ -\frac{5}{6} & \frac{4}{9} \end{pmatrix} = \begin{pmatrix} 1 & 0 \\ 0 & 1 \end{pmatrix}$,

and $BA = \begin{pmatrix} \frac{6}{9} & -\frac{3}{9} \\ -\frac{5}{9} & \frac{4}{9} \end{pmatrix}\begin{pmatrix} 4 & 3 \\ 5 & 6 \end{pmatrix} = \begin{pmatrix} 1 & 0 \\ 0 & 1 \end{pmatrix}$.

Since $AB = BA = I$, we say that A is the inverse of B, or B is the inverse of A.

$\therefore A = B^{-1}$ or $B = A^{-1}$.

In general, if $A = \begin{pmatrix} a & b \\ c & d \end{pmatrix}$

then $\qquad A^{-1} = \begin{pmatrix} \dfrac{d}{\Delta} & \dfrac{-b}{\Delta} \\ \dfrac{-c}{\Delta} & \dfrac{a}{\Delta} \end{pmatrix}$ where $\Delta = ad - bc$.

Δ is called the *determinant* of the matrix.

If $\Delta = 0$, then the matrix will have no inverse, and is called a *singular* matrix. A matrix which has an inverse is called *non-singular*.

Find the inverse of the following matrices where possible:

1 $\begin{pmatrix} 4 & 2 \\ 5 & 3 \end{pmatrix}$ 　　 2 $\begin{pmatrix} 6 & -7 \\ -5 & 6 \end{pmatrix}$ 　　 3 $\begin{pmatrix} 4 & 2 \\ -8 & -4 \end{pmatrix}$

4 $\begin{pmatrix} 2 & 3 \\ -6 & 9 \end{pmatrix}$ 　　 5 $\begin{pmatrix} 1 & 0 \\ 0 & -1 \end{pmatrix}$ 　　 6 $\begin{pmatrix} 6 & -4 \\ 3 & 11 \end{pmatrix}$

9.7　Algebraic Manipulation 2

Because multiplication of matrices is not commutative, care has to be taken in manipulating matrix equations.

For example,

$$\text{if } AC = B, \text{ where } A, B \text{ and } C \text{ are matrices,}$$

multiply both sides on the *left* by A^{-1}:

$$\therefore A^{-1}AC = A^{-1}B,$$

$A^{-1}A = I,$ 　　　 $\therefore$ 　　 $IC = A^{-1}B,$

$$\therefore \quad C = A^{-1}B. \quad \text{(Provided } A^{-1}B \text{ is possible.)}$$

9.8　Simultaneous Equations

Consider the equations 　　　 $2x + 4y = 12$ 　　　　　　 (1)

$$3x - 7y = 5. \tag{2}$$

These can be represented using matrices in the following way:

$$\begin{pmatrix} 2 & 4 \\ 3 & -7 \end{pmatrix} \begin{pmatrix} x \\ y \end{pmatrix} = \begin{pmatrix} 12 \\ 5 \end{pmatrix}.$$

If $A = \begin{pmatrix} 2 & 4 \\ 3 & -7 \end{pmatrix}$, $B = \begin{pmatrix} 12 \\ 5 \end{pmatrix}$ and $C = \begin{pmatrix} x \\ y \end{pmatrix}$,

then we have $AC = B$,
as in section 9.6 this means $C = A^{-1}B$.

$$A^{-1} = \begin{pmatrix} \frac{-7}{-26} & \frac{-4}{-26} \\ \frac{-3}{-26} & \frac{2}{-26} \end{pmatrix} = \begin{pmatrix} \frac{7}{26} & \frac{2}{13} \\ \frac{3}{26} & \frac{-1}{13} \end{pmatrix},$$

$$\therefore \begin{pmatrix} x \\ y \end{pmatrix} = \begin{pmatrix} \frac{7}{26} & \frac{2}{13} \\ \frac{3}{26} & \frac{-1}{13} \end{pmatrix} \begin{pmatrix} 12 \\ 5 \end{pmatrix} = \begin{pmatrix} 4 \\ 1 \end{pmatrix};$$

the solution is $x = 4, y = 1$.

Examination questions often slightly alter the wording as shown in the following example:

Worked Example 9.1

(a) Find $\begin{pmatrix} 4 & 3 \\ -1 & 9 \end{pmatrix} \begin{pmatrix} 9 & -3 \\ 1 & 4 \end{pmatrix}$.

(b) Use your answer to part (a) to solve the equations

$$4t + 3q = 31, \qquad 9q - t = 41.$$

Solution

(a) $\begin{pmatrix} 4 & 3 \\ -1 & 9 \end{pmatrix} \begin{pmatrix} 9 & -3 \\ 1 & 4 \end{pmatrix} = \begin{pmatrix} 39 & 0 \\ 0 & 39 \end{pmatrix}$

(b) The first part of the question in fact gives us straight away the determinant of the matrix as 39. Notice that the equations have been muddled up, and in fact are

$$4t + 3q = 31,$$

$$-t + 9q = 41.$$

The relevance of the matrices can now be seen.

Find $\begin{pmatrix} 9 & -3 \\ 1 & 4 \end{pmatrix} \begin{pmatrix} 31 \\ 41 \end{pmatrix} = \begin{pmatrix} 156 \\ 195 \end{pmatrix}$.

This answer must be divided by 39.

$\therefore t = 4, q = 5$.

This method is slightly easier in that it does not involve any fractions.

If $X = \begin{pmatrix} a & b \\ c & d \end{pmatrix}$, the matrix $\begin{pmatrix} d & -b \\ -c & a \end{pmatrix}$ is called the *adjoint* of X.

Check 26

1 $A = \begin{pmatrix} 1 & 4 \\ -1 & 3 \end{pmatrix}$, $B = \begin{pmatrix} 2 & -2 \\ 4 & 1 \end{pmatrix}$, $C = \begin{pmatrix} 3 & -1 & -1 \\ 0 & 1 & 2 \end{pmatrix}$,

$D = \begin{pmatrix} 4 & 3 & 0 \\ -2 & -1 & 1 \end{pmatrix}$, $E = \begin{pmatrix} 2 & 1 \\ 4 & -2 \\ 3 & -5 \end{pmatrix}$, $F = \begin{pmatrix} 4 & 3 \\ -1 & 0 \\ 1 & 2 \end{pmatrix}$ and $G = \begin{pmatrix} 1 & -1 & -2 \\ 0 & 1 & 3 \\ 4 & 1 & 6 \end{pmatrix}$.

Evaluate if possible:
(a) $C + 2D$;   (b) $3E - 2F$;   (c) $A - 3B$;   (d) $2D - 3E$;
(e) $4G$;   (f) $A + B + 2C$.

2 Find the matrix M, given that:

$$4M - \begin{pmatrix} 1 & 2 \\ -3 & 5 \end{pmatrix} = \begin{pmatrix} 6 & 1 \\ -1 & 4 \end{pmatrix}.$$

3 Find the matrix A if

$$\begin{pmatrix} 1 & 0 \\ 2 & -1 \end{pmatrix} - 3A = 2A + \begin{pmatrix} -1 & 0 \\ 1 & 2 \end{pmatrix}.$$

143

4 Evaluate, if possible, the following matrix products:

(a) $(2 \quad 1)\begin{pmatrix} 3 \\ 4 \end{pmatrix}$ (b) $\begin{pmatrix} 3 \\ 1 \end{pmatrix}(2 \quad 1)$ (c) $(1 \quad 3 \quad 4)\begin{pmatrix} 2 \\ -1 \\ -2 \end{pmatrix}$

(d) $(2 \quad 1)\begin{pmatrix} 3 \\ 4 \\ 2 \end{pmatrix}$ (e) $\begin{pmatrix} 3 & 1 & -1 \\ 0 & 1 & 1 \end{pmatrix}\begin{pmatrix} 2 & 1 \\ 4 & 2 \\ 1 & 0 \end{pmatrix}$

(f) $\begin{pmatrix} -1 & 0 \\ 2 & -3 \end{pmatrix}\begin{pmatrix} 4 \\ 1 \end{pmatrix}$ (g) $\begin{pmatrix} 1 & -1 & 0 \\ 2 & 1 & 3 \end{pmatrix}\begin{pmatrix} 2 \\ -1 \\ 0 \end{pmatrix}$.

(h) $\begin{pmatrix} 2 & 1 \\ 1 & 2 \end{pmatrix}\begin{pmatrix} 3 & 4 & -1 \\ 0 & 1 & 2 \end{pmatrix}$ (i) $\begin{pmatrix} 2 & 2 & 0 \\ 1 & 3 & 4 \\ -1 & 0 & 6 \end{pmatrix}\begin{pmatrix} 1 & 4 & -2 \\ 0 & 1 & 3 \\ 3 & 1 & 6 \end{pmatrix}$

5 If $X = \begin{pmatrix} 2 & 1 \\ 3 & 0 \end{pmatrix}$, $Y = \begin{pmatrix} -1 & 0 \\ 2 & 4 \end{pmatrix}$ and $Z = \begin{pmatrix} 3 & 1 \\ -1 & 0 \end{pmatrix}$, find XYZ.

6 Solve the following pairs of simultaneous equations by using a matrix method.

(a) $3x + 4y = 1$
$ 2x - y = 6$
(b) $5x + 3y = 7$
$ 4x - 5y = 9$
(c) $6x = 5y - 11$
$ 2y = 8x + 3$
(d) $6x - 9y = 11$
$ 5x + 17y = 2$

9.9   Networks and Route Matrices

Figure 9.1 shows a network consisting of four *nodes*, A, B, C and D, linked by seven *arcs* in various ways, which divide the plane up into five *regions*. The routes represented by the arcs can be travelled along in either direction.

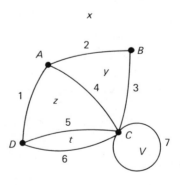

Fig. 9.1

(a) One-stage Route Matrix

A *one-stage* route between any two nodes is not allowed to pass through any of the other nodes. If R is the one-stage route matrix, then

$$
R = \text{from} \begin{array}{c} \\ A \\ B \\ C \\ D \end{array} \overset{\overset{\text{to}}{\begin{array}{cccc} A & B & C & D \end{array}}}{\begin{pmatrix} 0 & 1 & 1 & 1 \\ 1 & 0 & 1 & 0 \\ 1 & 1 & 2 & 2 \\ 1 & 0 & 2 & 0 \end{pmatrix}}.
$$

Note that the matrix is *symmetrical* about the leading diagonal.

(b) Two-stage Route Matrix

A *two-stage* route between two nodes passes through one other node on the way. The two-stage routes between A and C are shown in Fig. 9.2. It can be seen that there are five.

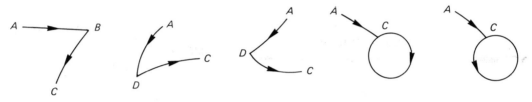

Fig. 9.2

If the reader now evaluates R^2, he will find that:

$$
R^2 = \begin{array}{c} \\ A \\ B \\ C \\ D \end{array} \overset{\begin{array}{cccc} A & B & C & D \end{array}}{\begin{pmatrix} 3 & 1 & 5 & 2 \\ 1 & 2 & 3 & 3 \\ 5 & 3 & 10 & 5 \\ 2 & 3 & 5 & 5 \end{pmatrix}}.
$$

This matrix gives the two-stage routes. The reader may like to try and justify ten for C to C. Similarly, R^3 gives the three-stage routes for the network.

9.10 Incidence Matrices

Let us look again at Fig. 9.1. Node A lies on (is *incident* on) arcs 1, 2, 4. Node C is incident on 3, 4, 5, 6 and 7 (twice). The matrix N representing the incidence of nodes on arcs is given below:

$$
N = \begin{array}{c} \\ A \\ B \\ C \\ D \end{array} \overset{\begin{array}{ccccccc} 1 & 2 & 3 & 4 & 5 & 6 & 7 \end{array}}{\begin{pmatrix} 1 & 1 & 0 & 1 & 0 & 0 & 0 \\ 0 & 1 & 1 & 0 & 0 & 0 & 0 \\ 0 & 0 & 1 & 1 & 1 & 1 & 2 \\ 1 & 0 & 0 & 0 & 1 & 1 & 0 \end{pmatrix}}.
$$

If we form the *transpose* N^T of the matrix (interchange rows and columns), we get

$$N^T = \begin{array}{c} \\ 1 \\ 2 \\ 3 \\ 4 \\ 5 \\ 6 \\ 7 \end{array} \begin{array}{cccc} A & B & C & D \\ \left(\begin{array}{cccc} 1 & 0 & 0 & 1 \\ 1 & 1 & 0 & 0 \\ 0 & 1 & 1 & 0 \\ 1 & 0 & 1 & 0 \\ 0 & 0 & 1 & 1 \\ 0 & 0 & 1 & 1 \\ 0 & 0 & 2 & 0 \end{array} \right) \end{array}.$$

This gives the incidence of the arcs on the nodes. The reader should look at NN^T, and compare it with the one-stage route matrix R.

Incidence matrices can also be written down for arcs on regions, and regions on arcs.

9.11 Euler's Theorem

If in any network, R is the number of regions, N the number of nodes and A the number of arcs, then

$$R + N = A + 2.$$

In Fig. 9.1, $\qquad\qquad\qquad\qquad 5 + 4 = 7 + 2.$

9.12 Traversability

If a network can be drawn without taking the pencil off the paper and without following the same arc twice, it is said to be *traversable*.

A network is traversable if
(a) all its junctions are even or (b) only two of its junctions are odd.

The *order* of a node is the number of arcs leaving the node.

In Fig. 9.3 the order of the nodes is as follows:

$$A(4), B(4), C(4), D(4), E(4).$$

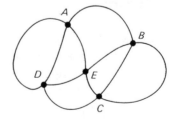

Fig. 9.3

Since these are all even, the network is traversable.

9.13   Transformations

It was stated in section 8.2 that 2×2 matrices can be used to represent transformations in the (x, y) plane. The simplest way to observe the effect of these is to consider the effect that a matrix has on the coordinates of a square with sides of length one, called the *unit square*.

(a) Reflection

The coordinates of a point (Fig. 9.4) are represented as a vector, e.g.

$$B = \begin{pmatrix} 1 \\ 1 \end{pmatrix}.$$

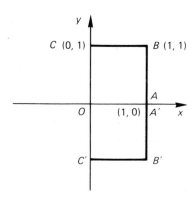

Fig. 9.4

The square can be represented as a 2×4 matrix:

$$\begin{array}{cccc} O & A & B & C \end{array}$$
$$\text{Square} = \begin{pmatrix} 0 & 1 & 1 & 0 \\ 0 & 0 & 1 & 1 \end{pmatrix}.$$

Consider the effect of the matrix $\begin{pmatrix} 1 & 0 \\ 0 & -1 \end{pmatrix}$. The square must be *premultiplied* by the matrix.

$$\begin{array}{cccc} & O' & A' & B' & C' \end{array}$$
$$\therefore \begin{pmatrix} 1 & 0 \\ 0 & -1 \end{pmatrix}\begin{pmatrix} 0 & 1 & 1 & 0 \\ 0 & 0 & 1 & 1 \end{pmatrix} = \begin{pmatrix} 0 & 1 & 1 & 0 \\ 0 & 0 & -1 & -1 \end{pmatrix}$$

Note: The first column of the matrix is the image of $\begin{pmatrix} 1 \\ 0 \end{pmatrix}$, the second of $\begin{pmatrix} 0 \\ 1 \end{pmatrix}$.

Transformed points will be denoted by A' etc. It can be seen that the vertices of the square have been reflected in the x-axis. It can also be shown as follows that all points will be reflected in the x-axis.

Consider any point $P(x, y)$, written $\begin{pmatrix} x \\ y \end{pmatrix}$. Transforming this point,

$$\begin{pmatrix} 1 & 0 \\ 0 & -1 \end{pmatrix}\begin{pmatrix} x \\ y \end{pmatrix} = \begin{pmatrix} x \\ -y \end{pmatrix}.$$

147

Hence $P(x, y) \rightarrow P'(x, -y)$. The x-coordinates has stayed the same, the y-coordinate has changed sign.

Similarly, $\begin{pmatrix} -1 & 0 \\ 0 & 1 \end{pmatrix}$ represents reflection in the y-axis,

$\begin{pmatrix} 0 & 1 \\ 1 & 0 \end{pmatrix}$ represents reflection in the line $y = x$,

$\begin{pmatrix} 0 & -1 \\ -1 & 0 \end{pmatrix}$ represents reflection in the line $y = -x$.

A very useful formula for finding the matrix which represents the transformation of reflection in the line $y = x \tan \alpha°$ ($\tan \alpha°$ is the gradient of the line which slopes at an angle $\alpha°$ to the x-axis) is given by

$$M = \begin{pmatrix} \cos 2\alpha° & \sin 2\alpha° \\ \sin 2\alpha° & -\cos 2\alpha° \end{pmatrix}.$$

For example, to find reflection in the line $y = x$, means that $\alpha = 45°$.

$$\therefore M = \begin{pmatrix} \cos 90° & \sin 90° \\ \sin 90° & -\cos 90° \end{pmatrix} = \begin{pmatrix} 0 & 1 \\ 1 & 0 \end{pmatrix}.$$

The origin $(0, 0)$ always stays at $(0, 0)$ under a 2×2 matrix transformation, and will be left out from now on. We say that $(0, 0)$ is *invariant*.

(b) Rotation

Consider the matrix $\begin{pmatrix} 0 & 1 \\ -1 & 0 \end{pmatrix}$;

$$\begin{matrix} & A & B & C & & A' & B' & C' \\ \begin{pmatrix} 0 & 1 \\ -1 & 0 \end{pmatrix} & \begin{pmatrix} 1 & 1 & 0 \\ 0 & 1 & 1 \end{pmatrix} & & = & \begin{pmatrix} 0 & 1 & 1 \\ -1 & -1 & 0 \end{pmatrix}. \end{matrix}$$

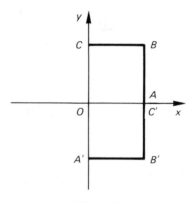

Fig. 9.5

The square appears to be in the same place (Fig. 9.5), but on looking at the labelling of the points, it can be seen that the square has been rotated *clockwise* by $90°$ about O, i.e. $-90°$. Clockwise angles are designated as being negative. Also, $\begin{pmatrix} 0 & -1 \\ 1 & 0 \end{pmatrix}$ is a rotation of $+90°$ about O, $\begin{pmatrix} -1 & 0 \\ 0 & -1 \end{pmatrix}$ is a half-turn about O.

In general, the matrix which gives an anticlockwise rotation of $\alpha°$ is given by

$$R = \begin{pmatrix} \cos \alpha° & -\sin \alpha° \\ \sin \alpha° & \cos \alpha° \end{pmatrix}.$$

(c) Shear

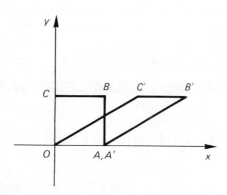

Fig. 9.6

Consider the effect of $\begin{pmatrix} 1 & 3 \\ 0 & 1 \end{pmatrix}$ on the unit square

$$\begin{pmatrix} 1 & 3 \\ 0 & 1 \end{pmatrix}\begin{pmatrix} 1 & 1 & 0 \\ 0 & 1 & 1 \end{pmatrix} = \begin{pmatrix} 1 & 4 & 3 \\ 0 & 1 & 1 \end{pmatrix}.$$

The square has been sheared parallel to the x-axis, the x-axis is the invariant line and $(1, 1) \rightarrow (4, 1)$.

In general, $\begin{pmatrix} 1 & k \\ 0 & 1 \end{pmatrix}$ is a shear of shear factor k parallel to the x-axis,

$\begin{pmatrix} 1 & 0 \\ k & 1 \end{pmatrix}$ is a shear of shear factor k parallel to the y-axis.

See section 8.7.

(d) Enlargement

If the matrix $\begin{pmatrix} k & 0 \\ 0 & k \end{pmatrix}$ is applied to the unit square, it simply enlarges it by a scale factor k, with O as the centre of the enlargement.
    See section 8.6.

(e) Stretch

If the matrix $\begin{pmatrix} k & 0 \\ 0 & 1 \end{pmatrix}$ is applied to the unit square, the square is stretched by a factor k in the direction of the x-axis. (See section 8.8.)

Similarly, $\begin{pmatrix} 1 & 0 \\ 0 & k \end{pmatrix}$ is a stretch parallel to the y-axis.

9.14   Combining Transformations

> If we want to find the effect of the transformation represented by matrix M_1, followed by the transformation represented by the matrix M_2, simply evaluate the matrix product $M_2 M_1$ (order must be correct).

9.15   Inverse and Identity Transformations

The matrix $\begin{pmatrix} 1 & 0 \\ 0 & 1 \end{pmatrix}$ as a transformation leaves the shape unchanged. For a transformation represented by T, the inverse matrix T^{-1} will represent the transformation that returns the shape to its original position.

Since many transformations are self-inverses, for example all reflections, the inverse of the matrix will be the same as the original matrix.

Worked Example 9.2

The matrix, X, gives information about the number of chocolates of each kind in three different boxes:

$$X = \begin{array}{c} \\ \text{Box } A \\ \text{Box } B \\ \text{Box } C \end{array} \begin{array}{c} \text{Nut} \\ \\ \left(\begin{array}{c} 3 \\ 4 \\ 6 \end{array} \right. \end{array} \begin{array}{c} \text{Soft} \\ \text{centre} \\ 5 \\ 3 \\ 6 \end{array} \begin{array}{c} \text{Hard} \\ \text{centre} \\ 4 \\ 5 \\ 3 \end{array} \begin{array}{c} \text{Toffee} \\ \\ 2 \\ 3 \\ 0 \end{array} \begin{array}{c} \text{Plain} \\ \\ 3 \\ 1 \\ 4 \end{array} \left. \begin{array}{c} \\ \\ \\ \\ \end{array} \right)$$

(a) Multiply matrix X by the matrix $\begin{pmatrix} 1 \\ 1 \\ 1 \\ 1 \\ 1 \end{pmatrix}$ and state what information your answer gives.

(b) Matrix Y gives information about the cost to the manufacturer of each type of chocolate and the number of calories in each type.

$$Y = \begin{array}{l} \text{Nut} \\ \text{Soft centre} \\ \text{Hard centre} \\ \text{Toffee} \\ \text{Plain} \end{array} \begin{array}{cc} \text{Cost} & \text{Calories} \\ \text{(pence)} & \\ \begin{pmatrix} 4 & 30 \\ 3 & 25 \\ 2 & 30 \\ 3 & 40 \\ 1 & 20 \end{pmatrix} \end{array}.$$

Evaluate the matrix product XY. Hence state the total number of calories in box A. What is the cost to the manufacturer of producing the chocolates in box C?

[OLE]

Solution

(a)

$$\begin{array}{l} A \\ B \\ C \end{array} \begin{pmatrix} 3 & 5 & 4 & 2 & 3 \\ 4 & 3 & 5 & 3 & 1 \\ 6 & 6 & 3 & 0 & 4 \end{pmatrix} \begin{pmatrix} 1 \\ 1 \\ 1 \\ 1 \\ 1 \end{pmatrix} = \begin{array}{l} A \\ B \\ C \end{array} \begin{pmatrix} 17 \\ 16 \\ 19 \end{pmatrix}$$

The answer gives the *total* number of chocolates in box A, box B and box C.

$$XY = \begin{array}{l} A \\ B \\ C \end{array} \begin{pmatrix} 3 & 5 & 4 & 2 & 3 \\ 4 & 3 & 5 & 3 & 1 \\ 6 & 6 & 3 & 0 & 4 \end{pmatrix} \begin{array}{cc} \text{Cost} & \text{Cal} \\ \begin{pmatrix} 4 & 30 \\ 3 & 25 \\ 2 & 30 \\ 3 & 40 \\ 1 & 20 \end{pmatrix} \end{array} = \begin{array}{l} A \\ B \\ C \end{array} \begin{array}{cc} \text{Cost} & \text{Cal} \\ \begin{pmatrix} 44 & 475 \\ 45 & 485 \\ 52 & 500 \end{pmatrix} \end{array}$$

It should be noted that the headings for the columns in the first matrix are the same as the headings for the rows in the second matrix. The headings for the final matrix are the row headings from the first matrix and the column headings from the second. Providing care is taken, interpreting the answer should not be difficult.

Total number of calories in box A is 475.

Total cost of producing box C is 52p.

Worked Example 9.3

Using a scale of 2 cm to represent 1 unit on each axis, draw on graph paper $\triangle PQR$, where P, Q, R are respectively the points $(3, -1)$ $(3, -2)$ and $(5, -2)$. Find and draw the image of $\triangle PQR$ under the transformation whose matrix is M where $M = \begin{pmatrix} 1 & 0 \\ 0 & -1 \end{pmatrix}$, and describe this transformation.

Find and draw the image of $\triangle PQR$ under the transformation whose matrix is NM, where

$$N = \begin{pmatrix} -\frac{3}{5} & \frac{4}{5} \\ \frac{4}{5} & \frac{3}{5} \end{pmatrix},$$

and describe the transformation whose matrix is N. By finding the coordinates of the point (x, y) under the transformation whose matrix is N, show that the line $y = 2x$ is unchanged under the transformation, and describe the transformation whose matrix is N.

[L]

Solution

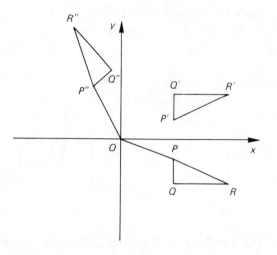

Fig. 9.7

Write the triangle as a 2 × 3 matrix:

$$\begin{array}{ccc} P & Q & R \end{array}$$
$$\Delta = \begin{pmatrix} 3 & 3 & 5 \\ -1 & -2 & -2 \end{pmatrix}.$$

$$\begin{array}{ccc} P' & Q' & R' \end{array}$$
$$\therefore \begin{pmatrix} 1 & 0 \\ 0 & -1 \end{pmatrix} \begin{pmatrix} 3 & 3 & 5 \\ -1 & -2 & -2 \end{pmatrix} = \begin{pmatrix} 3 & 3 & 5 \\ 1 & 2 & 2 \end{pmatrix}.$$

This is reflection in the x-axis.

$$NM = \begin{pmatrix} -\frac{3}{5} & \frac{4}{5} \\ \frac{4}{5} & \frac{3}{5} \end{pmatrix} \begin{pmatrix} 1 & 0 \\ 0 & -1 \end{pmatrix} = \begin{pmatrix} -\frac{3}{5} & -\frac{4}{5} \\ \frac{4}{5} & -\frac{3}{5} \end{pmatrix},$$

$$\begin{array}{ccc} P'' & Q'' & R'' \end{array}$$
$$\begin{pmatrix} -\frac{3}{5} & -\frac{4}{5} \\ \frac{4}{5} & -\frac{3}{5} \end{pmatrix} \begin{pmatrix} 3 & 3 & 5 \\ -1 & -2 & -2 \end{pmatrix} = \begin{pmatrix} -1 & -\frac{1}{5} & -\frac{7}{5} \\ 3 & \frac{18}{5} & \frac{26}{5} \end{pmatrix}.$$

The triangle has been rotated about O anticlockwise by an angle of $126.9°$.
   Transforming (x, y) by N we have

$$\begin{pmatrix} -\frac{3}{5} & \frac{4}{5} \\ \frac{4}{5} & \frac{3}{5} \end{pmatrix} \begin{pmatrix} x \\ y \end{pmatrix} = \begin{pmatrix} -\frac{3}{5}x + \frac{4}{5}y \\ \frac{4}{5}x + \frac{3}{5}y \end{pmatrix};$$

if $y = 2x$, the matrix on the right becomes

$$\begin{pmatrix} -\frac{3}{5}x + \frac{8}{5}x \\ \frac{4}{5}x + \frac{6}{5}x \end{pmatrix} = \begin{pmatrix} x \\ 2x \end{pmatrix},$$

hence y is still equal to $2x$.

If a complete line is unaltered, unless some shearing has occurred, which the diagram shows it hasn't, the transformation N is a reflection in the line $y = 2x$.

Worked Example 9.4

(a) Given that $A = \begin{pmatrix} 2 & 4 \\ 3 & 1 \end{pmatrix}$, $I = \begin{pmatrix} 1 & 0 \\ 0 & 1 \end{pmatrix}$ and k is a number, write down $A - kI$ as a simple matrix. Denote this matrix by $X(k)$.

(b) Write down the determinant of $X(k)$ and show that $X(k)$ has no inverse if $k^2 - 3k - 10 = 0$.

   If k_1 and k_2 are the solutions of this equation, with $k_1 > k_2$, find k_1 and k_2. Hence write down the matrices $P = X(k_1)$ and $Q = X(k_2)$ and evaluate the matrix product PQ.

(c) Find the images of A $(0, 2)$ and B $(2, 0)$ under the transformation represented by Q. Write down the equation of the line AB and show that every point on AB has the same image under this transformation. [AEB]

Solution

(a) $X(k) = \begin{pmatrix} 2 & 4 \\ 3 & 1 \end{pmatrix} - \begin{pmatrix} k & 0 \\ 0 & k \end{pmatrix} = \begin{pmatrix} 2-k & 4 \\ 3 & 1-k \end{pmatrix}$.

(b) The determinant $= (2-k)(1-k) - 3 \times 4$
$$= k^2 - 3k + 2 - 12 = k^2 - 3k - 10.$$

If the matrix has no inverse, the determinant is zero.
Hence $k^2 - 3k - 10 = 0$
$\therefore (k-5)(k+2) = 0$.
Since $k_1 > k_2$, $k_1 = 5$, $k_2 = -2$.

$$\therefore P = \begin{pmatrix} -3 & 4 \\ 3 & -4 \end{pmatrix}, \quad Q = \begin{pmatrix} 4 & 4 \\ 3 & 3 \end{pmatrix},$$

$$PQ = \begin{pmatrix} -3 & 4 \\ 3 & -4 \end{pmatrix}\begin{pmatrix} 4 & 4 \\ 3 & 3 \end{pmatrix} = \begin{pmatrix} 0 & 0 \\ 0 & 0 \end{pmatrix}.$$

(c)
$$\begin{array}{cccc} & A & B & A' \; B' \end{array}$$
$$\begin{pmatrix} 4 & 4 \\ 3 & 3 \end{pmatrix}\begin{pmatrix} 0 & 2 \\ 2 & 0 \end{pmatrix} = \begin{pmatrix} 8 & 8 \\ 6 & 6 \end{pmatrix}$$

hence $A \rightarrow (8, 6)$ and $B \rightarrow (8, 6)$.

To find what happens to *any* point on AB, we need the equation of AB, which is $x + y = 2$.

Hence any point (x, y) can be written $(x, 2 - x)$,

$$\begin{pmatrix} 4 & 4 \\ 3 & 3 \end{pmatrix}\begin{pmatrix} x \\ 2-x \end{pmatrix} = \begin{pmatrix} 4x + 8 - 4x \\ 3x + 6 - 3x \end{pmatrix} = \begin{pmatrix} 8 \\ 6 \end{pmatrix}.$$

Hence any point on AB is mapped onto $(8, 6)$.

Exercise 9

1 The inverse of the matrix $\begin{pmatrix} 1 & 4 \\ 2 & 7 \end{pmatrix}$ is:

  A $\begin{pmatrix} -7 & 4 \\ 2 & -1 \end{pmatrix}$ B $\begin{pmatrix} 7 & -4 \\ -2 & 1 \end{pmatrix}$ C $\begin{pmatrix} 1 & 2 \\ 4 & 7 \end{pmatrix}$ D $\begin{pmatrix} -1 & -2 \\ -4 & -7 \end{pmatrix}$

[AEB]

2 If $A = \begin{pmatrix} 1 & 1 \\ 1 & 1 \end{pmatrix}$, then A^2 equals:

  A $\begin{pmatrix} 1 & 1 \\ 1 & 1 \end{pmatrix}$ B $\begin{pmatrix} 1 & 0 \\ 0 & 1 \end{pmatrix}$ C $\begin{pmatrix} 2 & 2 \\ 2 & 2 \end{pmatrix}$ D $\begin{pmatrix} 0 & 1 \\ 1 & 0 \end{pmatrix}$

3 The matrix $\begin{pmatrix} 1 & 4 \\ -4 & 1 \end{pmatrix}$ represents:

  A   a rotation   B   a reflection   C   a shear   D   none of these

4 If $AB = \begin{pmatrix} 1 & 0 \\ 0 & 1 \end{pmatrix}$ where A and B are 2×2 matrices, which of the following could be false?

  A $A^2 B^2 = \begin{pmatrix} 1 & 0 \\ 0 & 1 \end{pmatrix}$ B $A = B$ C $(AB)^2 = \begin{pmatrix} 1 & 0 \\ 0 & 1 \end{pmatrix}$ D $A = B^{-1}$

5 If $A = \begin{pmatrix} 2 & 3 \\ -4 & -6 \end{pmatrix}$ and $A^2 = kA$ where k is a number, then k equals:

  A -4 B   4   C   1   D A

6 If $A = \begin{pmatrix} 4 & 3 \\ 1 & 1 \end{pmatrix}$, $B = \begin{pmatrix} 2 & 0 \\ 2 & 0 \end{pmatrix}$ and $C = (2 \quad 4)$, then $A + BC$ equals:

  A $\begin{pmatrix} 8 & 3 \\ 5 & 1 \end{pmatrix}$ B $\begin{pmatrix} 12 & 3 \\ 9 & 1 \end{pmatrix}$ C $\begin{pmatrix} 12 & 12 \\ 6 & 4 \end{pmatrix}$ D   none of these

7 The matrix which represents a reflection in the line $y = -x$, followed by an enlargement, centre O, scale factor 2, is:

  A $\begin{pmatrix} -2 & 0 \\ 0 & 2 \end{pmatrix}$ B $\begin{pmatrix} 2 & 0 \\ 0 & -2 \end{pmatrix}$ C $\begin{pmatrix} 0 & 2 \\ 2 & 0 \end{pmatrix}$ D $\begin{pmatrix} 0 & -2 \\ -2 & 0 \end{pmatrix}$

8 The inverse of AB where A and B are 2×2 matrices, whose inverses exist, is always:

  A $B^{-1}A^{-1}$ B AB^{-1} C $A^{-1}B$ D $A^{-1}B^{-1}$

9 If $A = \begin{pmatrix} a & b \\ c & d \end{pmatrix}$ and $B = \begin{pmatrix} -c & -d \\ a & b \end{pmatrix}$, then AB equals:

  A $\begin{pmatrix} 0 & 0 \\ 0 & 0 \end{pmatrix}$ B $\begin{pmatrix} -ac & -bd \\ ac & bd \end{pmatrix}$ C $\begin{pmatrix} ac + ab & -ad + b^2 \\ c^2 + da & bd - dc \end{pmatrix}$

  D   none of these

10 The transformation $T = \begin{pmatrix} 1 & 2 \\ 1 & -1 \end{pmatrix}$ maps P onto Q. If P lies on the line $x + 2y = 0$, then Q lies on the line:

  A $y = \dfrac{3x}{2}$ B $y = 2x$ C $x = 0$ D   none of these

11 If x and y are positive, and

$$\begin{pmatrix} x & x \\ y & y \end{pmatrix}\begin{pmatrix} x & y \\ y & x \end{pmatrix} = \begin{pmatrix} 3 & t \\ t & 3 \end{pmatrix},$$

find x, y and t.

12 If $A = (1 \quad 2)$ and $B = \begin{pmatrix} 3 \\ -1 \end{pmatrix}$, evaluate:

(a) AB (b) BA

13 If $A = \begin{pmatrix} x_1 & x_2 \\ x_3 & x_4 \end{pmatrix}$, $B = \begin{pmatrix} y_1 & y_2 \\ y_3 & y_4 \end{pmatrix}$ and $C = \begin{pmatrix} t_1 & t_2 \\ t_3 & t_4 \end{pmatrix}$

show that $A(BC) = (AB)C$.

14 Find x and y given that $\begin{pmatrix} x \\ y \end{pmatrix} = \begin{pmatrix} 4 & -1 \\ 2 & 0 \end{pmatrix}\begin{pmatrix} 3 \\ -2 \end{pmatrix} + \begin{pmatrix} 1 \\ 6 \end{pmatrix}$.

15 Given that $X = \begin{pmatrix} -1 & 0 \\ 0 & 1 \end{pmatrix}$, $P = \begin{pmatrix} 1 & 2 \\ 0 & 1 \end{pmatrix}$,

(a) calculate $X + P$;

(b) calculate XP;

(c) name the transformation represented by X. [AEB]

16 The following statements about the 2×2 matrices A, B, C and D are all false. Give examples which demonstrate these statements are false.

(a) $\begin{pmatrix} 1 & 0 \\ 0 & 0 \end{pmatrix}A = \begin{pmatrix} 0 & 0 \\ 0 & 0 \end{pmatrix} \Rightarrow A = \begin{pmatrix} 0 & 0 \\ 0 & 0 \end{pmatrix}$.

(b) $\begin{pmatrix} 1 & 0 \\ 0 & 0 \end{pmatrix}B = B\begin{pmatrix} 1 & 0 \\ 0 & 0 \end{pmatrix}$ for every 2×2 matrix B.

(c) If $\begin{pmatrix} 1 & 0 \\ 0 & 0 \end{pmatrix}C = \begin{pmatrix} 1 & 0 \\ 0 & 0 \end{pmatrix}D$, then $C = D$. [OLE]

17 If $A = \begin{pmatrix} 5 & 8 \\ -3 & -5 \end{pmatrix}$, evaluate A^2, and hence solve the equation

$5x + 8y = 2,$
$-3x - 5y = 1.$

18 A triangle is formed by the points A $(2, 0)$, B $(5, 0)$ and C $(4, 2)$. The matrix $\begin{pmatrix} 2 & 4 \\ 1 & 6 \end{pmatrix}$ transforms the triangle into the triangle $A'B'C'$. Find the determinant of the matrix, and hence find the area of triangle $A'B'C'$.

19 On graph paper, plot the square A $(2, 0)$, B $(2, 2)$, C $(0, 2)$ and D $(0, 0)$. Transform this square by means of the matrix $\begin{pmatrix} 0.6 & 0.8 \\ -0.8 & 0.6 \end{pmatrix}$, and hence give a geometrical description of the matrix.

20 Under a certain transformation T, the image (x', y') of the point (x, y) is given by:

$$\begin{pmatrix} x' \\ y' \end{pmatrix} = \begin{pmatrix} 4 & 2 \\ 0 & 3 \end{pmatrix}\begin{pmatrix} x \\ y \end{pmatrix} + \begin{pmatrix} 2 \\ 1 \end{pmatrix}.$$

(a) Find the image under T of $(0, 0)$.

(b) Find the point which under T is mapped onto $(16, 0)$.

(c) Is there a point P such that $T(P) = P$?

21 The following list defines a number of transformations:

R is a rotation of $90°$ anticlockwise about the origin.

H is a rotation of $180°$ about the origin.

M_x is a reflection in the x-axis.

M_y is a reflection in the y-axis.

T_x is a translation of two units parallel to the x-axis.

T_y is a translation of two units parallel to the y-axis.

The ends A and B of a line segment have coordinates $(2, 0)$ and $(2, 4)$ respectively. Show, by clear diagram, that the effect of the transformation $T_x M_y T_y R$ is to move AB to a position BC. State the coordinates of C.

Show also that $RT_y M_y T$ produces a different result but that $kRT_y M_y T_x$ will transform AB to BC if k is one of the transformations in the list above. Which transformation is k? [OLE]

22 ABC is a triangle whose sides are all of different lengths; D is the mid-point of BC, E is the mid-point of CA and F is the mid-point of AB. Half-turns with centres D, E and F are denoted by **D**, **E** and **F** respectively. The image of the point A under **D** is denoted by **D**(A), etc., and the image of the whole triangle is denoted by **D**$(\triangle)$.

Draw a sketch showing in a single diagram the triangle ABC and its images under **D**, **E** and **F**. State as fully as you can:

(a) the relationship between the points **D**(A), B and **F**(C);

(b) the relationship between the line segment CA and the line joining **D**(A) to **F**(C);

(c) the transformation mapping **D**$(\triangle)$ onto **F**$(\triangle)$. State the relationship between the line segments FD and AC. State also the type of transformation mapping the triangle DEF onto the triangle formed by the points **D**(A), **E**(B), **F**(C). [OLE]

23 If $\begin{pmatrix} 3 & 2 & -1 \\ 0 & 1 & 3 \\ -1 & 2 & 5 \end{pmatrix} \begin{pmatrix} a \\ 0 \\ b \end{pmatrix} = \begin{pmatrix} 4 \\ -3 \\ -6 \end{pmatrix}$, find a and b.

24 If $H = \begin{pmatrix} d \\ e \end{pmatrix}$ and $A = \begin{pmatrix} d & e^2 \\ -e & d^2 \end{pmatrix}$, evaluate:

$H^T A H$ (where H^T denotes the transpose of H). Deduce that

$(0 \quad x) \begin{pmatrix} 0 & x^2 \\ -x & 0 \end{pmatrix} \begin{pmatrix} 0 \\ x \end{pmatrix} = (0)$ whatever the value of x.

25 What is the matrix A which will transform the shape in Fig. 9.8(a) to that in Fig. 9.8(b)?

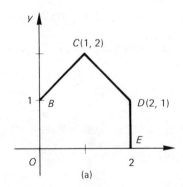

(a)

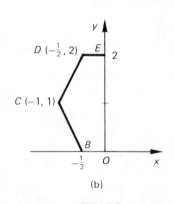

(b)

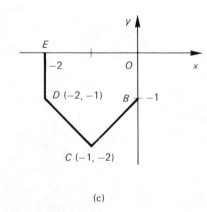

(c)

Fig. 9.8

Draw the figure obtained if the transformation A^2 is applied to Fig. 9.8(a).

Find the matrix B which transforms Fig. 9.8(a) into Fig. 9.8(c). What is the effect of applying B^n to Fig. 9.8(a), (a) if n is a positive integer, (b) if n is a negative integer?

26 Draw a sketch of the circle $x^2 + y^2 = 1$ after it has been transformed by the matrix $\begin{pmatrix} 2 & 0 \\ 0 & 1 \end{pmatrix}$.

27 (a) For the network shown in Fig. 9.9 write down:
    (i) the matrix X giving the number of arcs between any two nodes;
    (ii) the 3×7 incidence matrix T for arcs on nodes;
    (iii) the corresponding incidence matrix S showing the incidence of nodes on arcs.
    (iv) What is the relationship between S and T?
  (b) Evaluate (i) TS, (ii) $TS - X$. What does $TS - X$ represent?

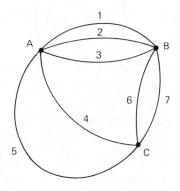

Fig. 9.9

28 (a) $A = \begin{pmatrix} 2 & 1 \\ -1 & 3 \end{pmatrix}$, $B = \begin{pmatrix} 4 & 1 \\ -2 & -5 \end{pmatrix}$, and $C = \begin{pmatrix} 3 & -1 \\ 2 & -2 \end{pmatrix}$.
Write, as a single matrix,
(i) $B + C$, (ii) AB.

  (b) $T = \begin{pmatrix} 0 & -1 \\ 1 & 0 \end{pmatrix}$. Evaluate the matrix product $T\begin{pmatrix} x \\ y \end{pmatrix}$ and describe clearly the transformation whose matrix is T. Hence, or otherwise, write T^8 as a single matrix.
                                                                   [UCLES]

29 If $A = \begin{pmatrix} 3 & 0 \\ 7 & 5 \end{pmatrix}$ and $B = \begin{pmatrix} 2 & 0 \\ 4 & t \end{pmatrix}$, find t given that $AB = 2A + 20\begin{pmatrix} 0 & 0 \\ 1 & 0 \end{pmatrix}$.

30 If $M = \begin{pmatrix} 1 & 1 \\ 1 & 2 \end{pmatrix}$ and $C = \begin{pmatrix} 1 \\ 1 \end{pmatrix}$, find the products MC and M^2C.

The product M^nC will be called $C_n = \begin{pmatrix} x_n \\ y_n \end{pmatrix}$. Write down x_1, y_1, x_2 and y_2 and find x_3 and y_3.

It is suggested that $y_n^2 = x_n x_{n+1} - 1$, for all values of n. Verify that the relation is true for two values of n already used, and find a similar expression for x_n^2 in terms of y_n and y_{n-1}.
Find a matrix N such that, for all values of n, $NC_n = C_{n-1}$.
Find a matrix P such that, for all values of n, $PC_n = C_{n+2}$.

31 The country of Albion has four local airstrips A, B, C, D, and two international airports G, H. Regular services exist or not between these, as shown by 1 or 0 in the matrix M. Flights from G and H to airports U, V, W in Gaul are given by matrix N.

Matrix $M = \begin{array}{c} \\ A \\ B \\ C \\ D \end{array}\begin{array}{cc} G & H \\ \left(\begin{array}{cc} 1 & 0 \\ 0 & 1 \\ 1 & 1 \\ 1 & 1 \end{array} \right) \end{array}$ Matrix $N = \begin{array}{c} \\ G \\ H \end{array}\begin{array}{ccc} U & V & W \\ \left(\begin{array}{ccc} 1 & 0 & 1 \\ 1 & 1 & 0 \end{array} \right) \end{array}$.

For the product matrix MN and interpret it. Local services in Gaul go from U, V, W to internal airports X, Y, Z as shown by matrix P.

Matrix $P = \begin{array}{c} \\ U \\ V \\ W \end{array}\begin{array}{ccc} X & Y & Z \\ \left(\begin{array}{ccc} 1 & 0 & 1 \\ 1 & 1 & 0 \\ 0 & 0 & 1 \end{array} \right) \end{array}$.

A traveller wishes to travel by air between these countries. Find by any method (a) which of the journeys from A or B or C or D to X or Y or Z have no air route; (b) which of them are connected by more than one route. [OLE]

32 (a) Under a certain transformation, the image (x', y') of a point (x, y) is given

by $\begin{pmatrix} x' \\ y' \end{pmatrix} = \begin{pmatrix} 2 & 1 \\ -1 & 1 \end{pmatrix}\begin{pmatrix} x \\ y \end{pmatrix} + \begin{pmatrix} 3 \\ 9 \end{pmatrix}$.

(i) Find the coordinates of A, the image of the point $(0, 0)$.
(ii) Find the coordinates of B, the image of the point $(4, 3)$.
(iii) Given that the image of the point (g, h) is the point $(0, 0)$, write down two equations each involving g and h. Hence, or otherwise, find the values of g and h.

(b) Find the value of x for which the matrix $\begin{pmatrix} x + 3 & 0 \\ 0 & 2 \end{pmatrix}$

(i) has no inverse;
(ii) represents an enlargement.

State the scale factor of this enlargement. [UCLES]

33 Using a scale of 1 cm to 1 unit on both axes, draw on graph paper the flag formed by the points $(2\ 1)$, $(2, 2)$, $(2, 3)$, $(3, 3)$ and $(3, 2)$. E is the enlargement with centre $(0, 2)$ and scale factor 2. T is the translation $\begin{pmatrix} -1 \\ -5 \end{pmatrix}$ which maps the point (x, y) onto the point $(x - 1, y - 5)$.

R is the anticlockwise rotation of $90°$ about the point $(3, -2)$.

On your graph, show the image of the flag under each of the following transformations:

(a) E; (b) T; (c) RT.

Label the images (i), (ii) and (iii) respectively.

The transformation RT is an anticlockwise rotation. State the angle of rotation and find the coordinates of the centre of the rotation. [UCLES]

10 Straight Lines and Circles

10.1   Parallel Lines

The basic properties of angles can be seen in Fig. 10.1.

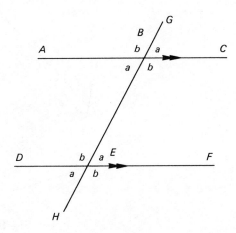

Fig. 10.1

BE can go in any direction; it is only the relative positions that are important.

The pairs ∠*ABG*, ∠*HBC* are called *vertically opposite* angles.
The pairs ∠*GBC*, ∠*BEF* are called *corresponding* angles.
The pairs ∠*ABE*, ∠*BEF* are called *alternate* angles; alternate angles form a characteristic 'Z' shape.
The pairs ∠*ABG*, ∠*GBC* are called *adjacent* angles.
Adjacent angles add up to 180°.
Any two angles that add up to 180° are called *supplementary* angles.

10.2   The Triangle

The angles of a triangle add up to 180°.
    The exterior angle of a triangle equals the sum of the two interior and opposite angles.

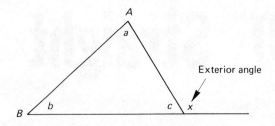

Fig. 10.2

In Fig. 10.2,

$$a + b + c = 180$$
$$a + b = x.$$

10.3 Polygons

The exterior angles of any polygon add up to 360°.

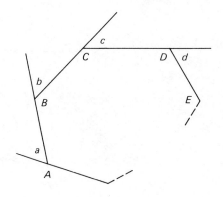

Fig. 10.3

Referring to Fig. 10.3,

$$a + b + c + d + \ldots = 360.$$

To investigate the sum of the interior angles of an n-sided polygon, we have:

$$\text{Sum} = (180 - a)° + (180 - b)° + (180 - c)° + (180 - d)° + \ldots$$
$$= (180 + 180 + 180 + 180 \ldots)° - (a + b + c + d + \ldots)°$$
$$= 180n - 360 = 90(2n - 4)°$$

We have: The sum of the interior angles of an n-sided polygon $= (2n - 4)$ right angles.

10.4 Regular Polygons

A *regular* n-sided polygon has n equal sides and n equal angles. Since the sum of the n equal exterior angles is 360°, it follows that for a regular polygon, each

$$\text{exterior angle} = \frac{360°}{n}.$$

10.5   Symmetry

(a) Line Symmetry

Consider the rectangle in Fig. 10.4. The line m divides the rectangle into two equal parts, one half of which can be folded about m to fit exactly on the other. We say that m is an *axis of symmetry* (or line symmetry).

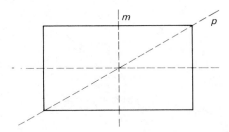

Fig. 10.4

Note that the line p divides the rectangle into two equal halves, but one half does not land on top of the other if folded about p, which is therefore not an axis of symmetry.

(b) Rotational Symmetry (Point Symmetry)

The hexagon in Fig. 10.5 maps onto itself if rotated about an axis through O at right angles to the plane by 60°. If this is repeated 6 times, A returns to its original position. We say that the *order* of rotational symmetry is 6.

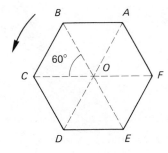

Fig. 10.5

(c) Three Dimensions

In three dimensions, we can have a *plane* of symmetry which divides a solid into two equal halves. For example, a plane through the centre of a sphere.

10.6 Geometric Properties of Simple Plane Figures

(a) Isosceles Triangle

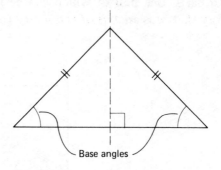

Fig. 10.6

Base angles are equal.

Two sides are equal.

Questions involving isosceles triangles are often made easier if considered as two joined right-angled triangles separated by the line of symmetry.

(b) Equilateral Triangle

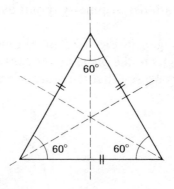

Fig. 10.7

All sides are equal.

All angles equal 60°.

There are 3 axes of symmetry.

Order of rotational symmetry is 3.

(c) Obtuse-angled Triangle

One angle is greater than 90°.

(d) Square

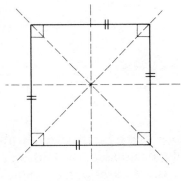

Fig. 10.8

All sides are equal.
All angles equal 90°.
Diagonals are equal in length and bisect each other at right angles.
There are 4 axes of reflective symmetry.
Order of rotational symmetry is 4.

(e) Rectangle

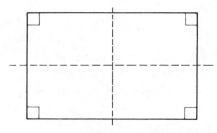

Fig. 10.9

All angles are 90°.
The diagonals are equal in length and bisect each other.
There are 2 axes of reflective symmetry.
Order of rotational symmetry is 2.

(b) Parallelogram

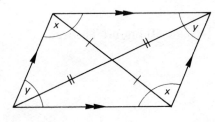

Fig. 10.10

Opposite sides are parallel.
Opposite angles are equal.
The diagonals bisect each other.
There is rotational symmetry of order 2 about the intersection of the diagonals.
There is no line symmetry.

163

(g) Rhombus

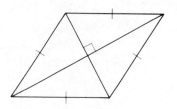

Fig. 10.11

A rhombus is a parallelogram with all sides equal.
The diagonals bisect each other at right angles.
There are two axes of line symmetry.
Order of rotational symmetry is 2.

(h) Kite

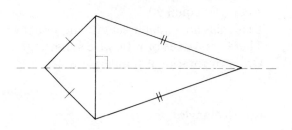

Fig. 10.12

The diagonals are perpendicular.
There is one axis of symmetry.
Two pairs of adjacent sides are equal in length.

(i) Trapezium

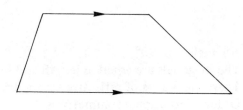

Fig. 10.13

Two opposite sides are parallel. In general, there is no symmetry.

(j) Triangle Mid-point Theorem

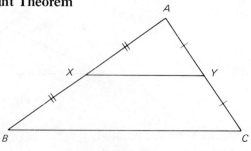

Fig. 10.14

In any triangle ABC, if the mid-points X and Y of AB and AC respectively are joined, XY is parallel to BC.

10.7 The Circle (Definitions)

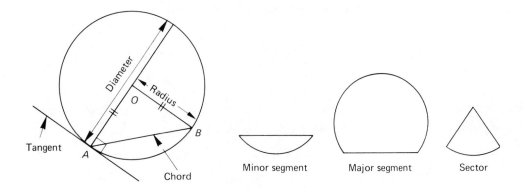

Fig. 10.15

The diameter is twice the radius. OAB is an isosceles triangle.

10.8 The Properties of the Circle

These will be stated here without proof.

(a) The angle between a radius OA and the tangent at A is 90°. See Fig. 10.15.

(b) The angle *subtended* at the centre by two points of the circle A and B is twice the angle subtended by A and B at any point T on the circle. See Fig. 10.16 for various configurations.

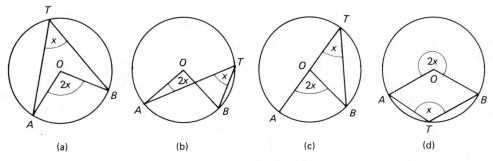

Fig. 10.16

(c) Angles in the same segment standing on the same chord are equal. Angles in opposite segments standing on the same chord are supplementary. See Fig. 10.17. AT_1BT_2 is called a *cyclic quadrilateral*.

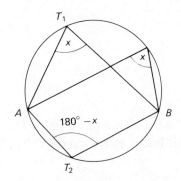

Fig. 10.17

(d) The angle between the tangent at T and a chord through A equals the angle on the chord in the opposite segment of the circle. See Fig. 10.18.

This property can be very useful.

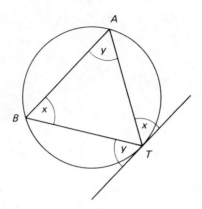

Fig. 10.18

(e) Tangents to a circle from a point are equal in length: In Fig. 10.19, the diagram is symmetrical about OT, hence $AT = BT$.

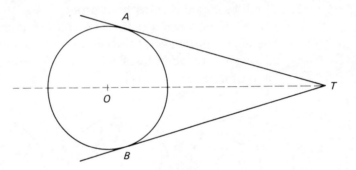

Fig. 10.19

(f) If two chords AB and CD intersect at P, it can be shown that $AP \cdot PB = PC \cdot PD$; see Fig. 10.20.

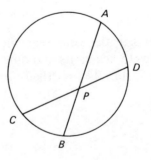

Fig. 10.20

(g) A line drawn from a point P outside a circle which cuts the circle at A and B (this line is called a *secant*) has the property

$$AB \cdot AC = AT^2$$

where T is the point of contact of the tangent from A.

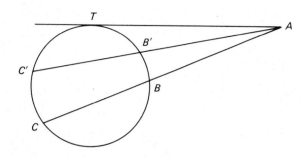

Fig. 10.21

For a second secant, $AB' \cdot AC' = AT^2$ (Fig. 10.21).

Hence, $$AB' \cdot AC' = AB \cdot AC.$$

Worked Example 10.1

A cyclic quadrilateral $PQRS$ has $PQ = PS$. The diagonal $QS = QR$. If $\angle QSR = 52°$, find $\angle PQR$. See Fig. 10.22.

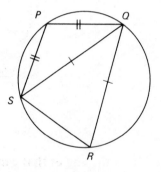

Fig. 10.22

Solution

Since $\triangle SQR$ is isosceles, $\angle SQR = 180° - 104° = 76°$.
$\angle SPQ$ and $\angle SRQ$ are in opposite segments,
hence $\angle SPQ = 180° - 52° = 128°$.
$\triangle SPQ$ is isosceles, $\therefore \angle PQS = \frac{1}{2}(180° - 128°) = 26°$,
$\therefore \angle PQR = 76° + 26° = 102°$.

Worked Example 10.2

Find the angles p and q marked in Fig. 10.23 where EA and ED are tangents to the circle.

Solution

First attempts at this question might give difficulty in using the recognized circle theorems. However, tangents drawn to a circle from a given point have the same

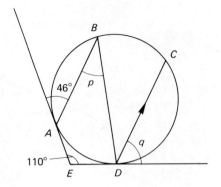

Fig. 10.23

length; see theorem (e) above. Hence, if the line *AD* is joined, since *EA* = *ED*, then *ADE* is an isosceles triangle.

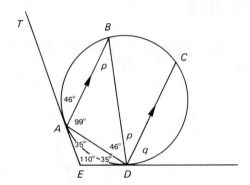

Fig. 10.24

In Fig. 10.24 the angles that can next be found are as follows:
$\angle EAD = \angle EDA = 35°$,
$\angle TAE = 180°, \therefore \angle BAD = 99°$.
$\angle BDA = \angle BAT = 46°$ (theorem (d)).
Hence $p = 35°$.
But $\angle BDC = \angle ABD = p$.
$\therefore 35° + 46° + p + q = 180°$,
hence $q = 64°$.

Worked Example 10.3

In Fig. 10.25, *PA* is 12 mm shorter than *PD*.

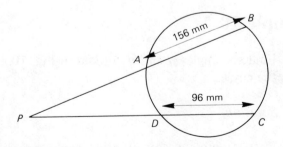

Fig. 10.25

It is also given that $AB = 156$ mm, $CD = 96$ mm and $PA = x$ mm.

(a) Obtain an equation for x.

(b) Calculate x. [AEB]

This is not an easy question to do unless the student remembers theorem (g).

(a) It follows that $PA \times PB = PD \times PC$.

$$\therefore x(x + 156) = (x + 12)(x + 12 + 96) = (x + 12)(x + 108).$$

(b) $$\therefore x^2 + 156x = x^2 + 120x + 1296,$$

$$\therefore 36x = 1296,$$

$$\therefore x = 36.$$

Check 27

Find all the angles marked with a letter in the following diagrams. T indicates a tangent.

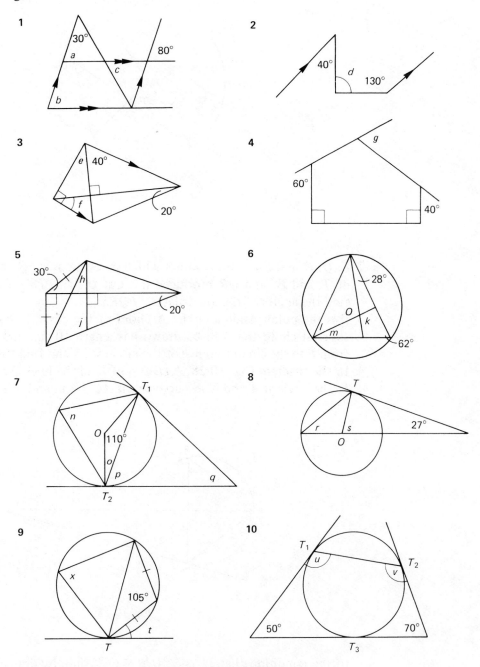

Fig. 10.26

Exercise 10

Multiple-choice questions on these topics are usually very straightforward and are therefore not included.

1. In Fig. 10.27, O is the centre of a circle of radius 8 cm. $\angle AOB = 80°$, $\angle CBO = 60°$, and BC and AO produced meet at D. Calculate:

   (a) $\angle ADB$ (b) AB (c) OD

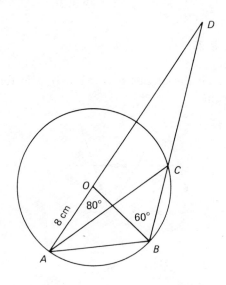

Fig. 10.27

2. P, Q, R and S are four points which lie on a circle PQ and SR produced meet at T, and PS and QR produced meet at U. If $\angle PTS = 30°$ and $\angle PUQ = 45°$, find the angles of the quadrilateral $PQRS$.

3. P is the point inside a circle. A chord of the circle is drawn through P. If the shortest chord that can be drawn has length 10 cm, and the greatest distance from P to the circumference of the circle is 25 cm, find the radius of the circle.

4. In the diagram Fig. 10.28, $\angle ADC = 30°$ and the lines DA and DC are tangents to the circle at A and C respectively. If $AC = 4$ m and B is able to move subject

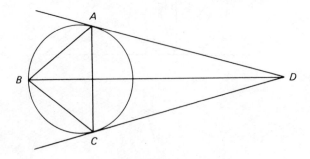

Fig. 10.28

to the condition that $30° \leqslant \angle ACB \leqslant 60°$, find, by drawing, the angle through which the line BD can turn.

5 In Fig. 10.29 *ABC* is a straight line, *O* is the centre of the circle and *AT* is a tangent to the circle. If $\angle BOT = 100°$ and $\angle BAT = 20°$, calculate $\angle TOC$.

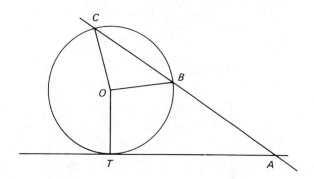

Fig. 10.29

6 In Fig. 10.30, *TA* and *TB* are tangents to a circle centre *O*, *TOC* is a straight line and $\angle ATB = 60°$. Calculate:
(a) $\angle ABT$ (b) $\angle AOB$ (reflex)   (c) $\angle BCO$

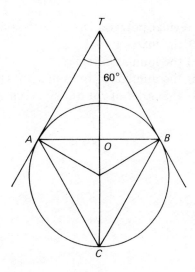

Fig. 10.30

7 In Fig. 10.31, *AC* is a diameter of the circle and the chord *BD* intersects *AC* at *X*. Given that $\angle BCA = 26°$ and $\angle CAD = 47°$, calculate:
(a) $\angle BAC$ (b) $\angle AXD$ [Cambridge]

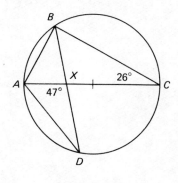

Fig. 10.31

8 *TA* is a tangent to a circle. (*A* is on the circle.) *AB* is a chord of the circle and *O* is the centre of the circle. If ∠*BAT* = 50°, find ∠*BOA*.

9 In Fig. 10.32, *O* is the centre of the circle and *BTC* is a tangent. Given that ∠*BAT* = *x*° and ∠*CAT* = 2*x*°, name three other angles of this figure each equal to 2*x*°.

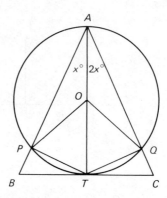

Fig. 10.32

10 A regular polygon has exterior angles of *x*°. Find, in terms of *x*,
   (a) the interior angle;
   (b) the number of sides;
   (c) the number of lines of symmetry;
   (d) the order of rotational symmetry of the polygon.

11 Probability and Statistics

11.1 Sample Space

The set of all possible outcomes in a given situation, or mathematical experiment, is called the sample space, denoted by S or $\mathcal{E}$.

Examples: (a) If a fair die is rolled,

$$S = \{1, 2, 3, 4, 5, 6\}.$$

(b) If three coins are tossed,

$$S = \{HHH, HHT, HTH, HTT, THH, THT, TTH, TTT\}.$$

This notation is not always very satisfactory; for example, if two dice are rolled, and the total score found, although there are only 12 scores, there are 36 possible outcomes. A much better way to illustrate the sample space would be by using a diagram such as that in Fig. 11.1.

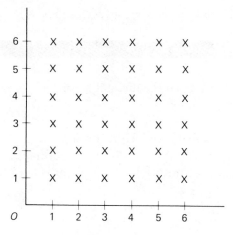

Fig. 11.1

11.2 Definition of Probability

(a) Empirical Definition

If an experiment is carried out n times and results in m successes, then the *empirical* (or experimental) probability is $\dfrac{m}{n}$.

If a die is rolled 18 times, and the score 2 is obtained twice, then the experimental probability is $\frac{2}{18} = \frac{1}{9}$. However, if asked what is the proability when rolling a die of scoring 2, we would probably give the answer of $\frac{1}{6}$. If we increase the number of times that the die is rolled, the probability would get closer to the theoretical value which is defined in the following section.

(b) Theoretical Definition

The theoretical definition is as follows:

$$\text{Probability} = \frac{\text{The number of ways of achieving success}}{\text{The total number of outcomes}}.$$

We have, then, that the theoretical probability of rolling a six with one die is $\frac{1}{6}$.

11.3 Complement, Certain and Impossible Events

The outcomes in a given situation are often referred to as events.

If an event is certain to happen, its probability is 1,
in fact $P(S) = 1$
If an event is impossible, its probability is zero, i.e. $P(\phi) = 0$.

Returning to the die and rolling a score of 2, we have

$$P(\text{score } 2) = \frac{1}{6}.$$

The probability of not scoring 2 is $\frac{5}{6}$.
The probability of an event not happening is called the *complement* of the event. Using set notation, if the event is denoted by E, its complement is denoted by E';

clearly, $\qquad\qquad\qquad\qquad P(E) + P(E') = 1.$

11.4 Set Notation

Questions in probability can often be made easier by the use of set notation. Consider the following example.

Worked Example 11.1

A card is taken from an ordinary pack of playing cards. Find in two different ways the probability of drawing a red picture card, illustrating one of them by a Venn diagram.

Solution

Method (a) The sample space S consists of the 52 cards. If the event of drawing a red picture card is denoted by R, then

$$R = \{J\heartsuit, Q\heartsuit, K\heartsuit, J\diamondsuit, Q\diamondsuit, K\diamondsuit\},$$

$$P(R) = \frac{n(R)}{n(S)} = \frac{6}{52} = \frac{3}{26}.$$

Method (b) Using the Venn diagram in Fig. 11.2,

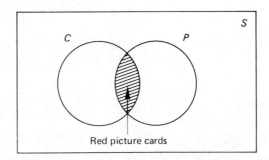

Fig. 11.2

If $C = \{\text{red cards}\}$ and $P = \{\text{picture cards}\}$, then the set of red picture cards is denoted by $C \cap P$.

$$P(R) = \frac{n(C \cap P)}{n(S)} = \frac{6}{52} = \frac{3}{26}.$$

The following check exercise consists of questions on basic probability.

Check 28

1 List all the possible outcomes if a coin is tossed 4 times.
2 A box contains 1 red, 1 white and 1 blue ball. Draw up a sample space for the possible outcomes when two balls are drawn (the first being replaced before the second is drawn).
3 A die is numbered 1, 1, 2, 5, 6, 6. Construct a sample space for rolling the die twice.
4 Determine the probabilities of the following events:
    (a) A king appears in drawing a single card from an ordinary pack of 52 cards.
    (b) A white marble appears in drawing a single marble from an urn containing 4 white, 3 red and 5 blue marbles.
    (c) Choosing a vowel from the word 'probability'.
    (d) A prime number is drawn from a pack of 25 cards numbered 1 to 25.
    (e) Not drawing a red ball from a box containing 6 red, 4 white and 5 blue balls.
5 $\mathscr{E} = \{1, 2, 3, \ldots, 50\}$, $A = \{\text{multiples of 7}\}$, $B = \{\text{multiples of 5}\}$, $C = \{\text{multiples of 3}\}$.
    If any element x is chosen at random from $\mathscr{E}$, find the probability that:
    (a) $x \in A \cup B$;   (b) $x \in B \cap C$;   (c) $x \in A \cap C'$.

11.5   Combined Events, Tree Diagrams

Consider the following situation. A fair coin is tossed, and a spinner numbered 1, 2, 3, 4 is spun and the results on each are recorded; the possible sample space S is given by

$$S = \{H1, H2, H3, H4, T1, T2, T3, T4\}.$$

We could also use a diagram similar to that in Fig. 11.1.

If E_1 is the event of gaining a head and a score of 2, then since $E_1 = \{H2\}$, we have

$$P(E_1) = \frac{n(E_1)}{n(S)} = \tfrac{1}{8}.$$

If E_2 is the event of gaining a head and an even number or a tail and scoring 1, then

$$E_2 = \{H2, H4, T1\},$$

hence $\qquad\qquad\qquad P(E_2) = \dfrac{n(E_2)}{n(S)} = \tfrac{3}{8}.$

We will now look at solving a problem of this nature by the use of a *tree diagram*, in which the probability of each event is marked on the appropriate branch of the tree. Since both tossing the coin and using the spinner are carried out at the same time, Fig. 11.3 shows the two possible tree diagrams that can be drawn. This is not the case if the order of the events matters, when only one can be drawn.

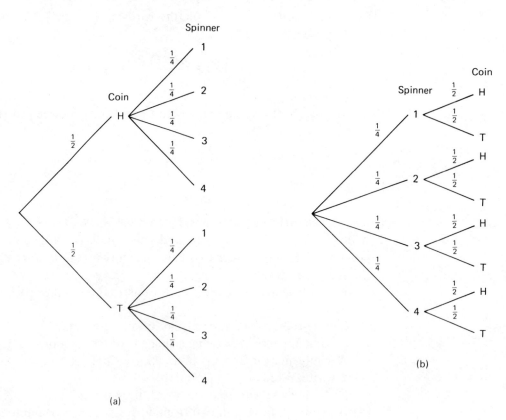

(a)   (b)

Fig. 11.3

To find $P(E_1)$: find the route on the branches of the tree which corresponds to H2 or 2H, and *multiply* the probabilities together of any branch you travel along:

$$P(E_1) = \tfrac{1}{2} \times \tfrac{1}{4} = \tfrac{1}{8} \text{ diagram (a)}$$

$$= \tfrac{1}{4} \times \tfrac{1}{2} = \tfrac{1}{8} \text{ diagram (b).}$$

The same answer is obtained either way.

To find $P(E_2)$: when there is more than one route that can be taken, you still multiply the probabilities on *each* route and *add* together the different answers.

In this case,

$$P(E_2) = \tfrac{1}{2} \times \tfrac{1}{4} + \tfrac{1}{2} \times \tfrac{1}{4} + \tfrac{1}{2} \times \tfrac{1}{4} = \tfrac{3}{8} \text{ diagram (a)}$$
$$= \tfrac{1}{4} \times \tfrac{1}{2} + \tfrac{1}{4} \times \tfrac{1}{2} + \tfrac{1}{4} \times \tfrac{1}{2} = \tfrac{3}{8} \text{ diagram (b).}$$

The following example considers the problem of the order of events.

Worked Example 11.2

Two cards are drawn one at a time from an ordinary pack of playing cards. Find the probability that they are both aces (a) if the first card is replaced, (b) if the first card is not replaced.

Solution

Although the probability trees are similar, the probabilities on the branches are not the same for the second card.

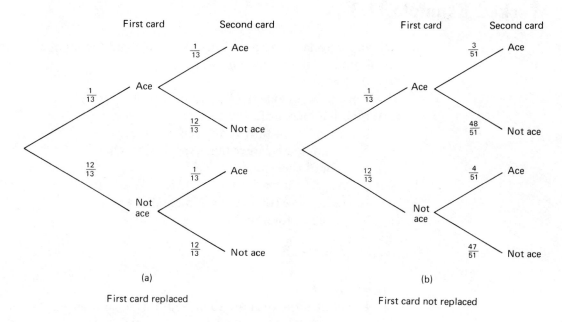

Fig. 11.4

(a) We have P (both aces) $= \tfrac{1}{13} \times \tfrac{1}{13} = \tfrac{1}{169}$.
(b) We have P (both aces) $= \tfrac{1}{13} \times \tfrac{3}{51} = \tfrac{1}{221}$.

The following exercise should be done using tree diagrams.

Check 29

1 Two dice are thrown. What is the probability of a score greater than 7?
2 Six discs, numbered 1 to 6, are placed in a bag. Two discs are drawn out

together. What is the probability that:

(a) the sum of the numbers on the discs is even;

(b) one of the discs has an odd number on it;

(c) at least one of the discs has an even number on it?

3 The probability of an event A happening is $\frac{1}{2}$ and the probability of an event B happening is $\frac{1}{4}$. Given that A and B are independent, calculate the probability that:

(a) neither event happens;

(b) just one of the two events happens.

4 A die numbered 1, 1, 2, 3, 4, 4 is thrown twice. Draw a sample space for the results. What is the probability that the total score on the two throws is (a) odd; (b) greater than 6?

5 A coin is biased in such a way that the probability of a head appearing in one toss of the coin is $\frac{2}{3}$. The coin is tossed three times. What is the probability that there are:

(a) three heads;

(b) at least two heads?

6 Given that the probability of a male birth is 0.52 and that a woman has three children, calculate the probability that:

(a) all three are boys;

(b) at least two are boys.

Worked Example 11.3

(a) Neil and Angela each roll a fair 6-faced die and the number on the top face is noted. What is the probability that:

(i) they each roll a 1;

(ii) Angela scores exactly 4 more than Neil;

(iii) their numbers differ by 3;

(iv) neither of them throws an even number;

(v) Neil's number is larger than Angela's number;

(vi) the sum of their numbers is 3?

(b) Neil and Angela each roll their die twice. What is the probability that:

(i) the total score (of all four numbers) is 5;

(ii) Neil scores a total of 2 and Angela a total of 3?   [OLE]

Solution

(a) (i) $\frac{1}{6} \times \frac{1}{6} = \frac{1}{36}$.

(ii) If Angela scores 4 more than Neil, this means Angela scores 56; write this A5 and N1 or A6 and N2.
∴ Probability $= \frac{1}{6} \times \frac{1}{6} + \frac{1}{6} \times \frac{1}{6} = \frac{1}{18}$.

(iii) (A6 and N3) or (A5 and N2) or (A4 and N1) or (A3 and N6) or (A2 and N5) or (A1 and N4).
∴ Probability $= 6 \times \frac{1}{6} \times \frac{1}{6} = \frac{1}{6}$.

(iv) If neither throws even, both throw odd (1, 3, 5).
∴ Probability $= \frac{1}{2} \times \frac{1}{2} = \frac{1}{4}$.

(v) If Neil's number is larger than Angela's number, possibilities are:
(N6, A5), (N6, A4), (N6, A3), (N6, A2), (N6, A1),
(N5, A4), (N5, A3), (N5, A2), (N5, A1), (N4, A3),
(N4, A2), (N4, A1), (N3, A2), (N3, A1), (N2, A1),
i.e. 15 possibilities.
∴ Probability $= 15 \times \frac{1}{6} \times \frac{1}{6} = \frac{5}{12}$.

(vi) If the sum is 3, we have
(N1, A2), (N2, A1).
$\therefore$ Probability $= 2 \times \frac{1}{6} \times \frac{1}{6} = \frac{1}{18}$.

(b) (i) The only possible scores on 4 dice to add up to 5 are $1 + 1 + 1 + 2$. The 2 could be on any of the 4 dice.
$\therefore$ Probability $= 4 \times (\frac{1}{6})^4 = \frac{1}{324}$.

(ii) The probability that Neil scores 2 is $\frac{1}{6} \times \frac{1}{6} = \frac{1}{36}$.
The probability that Angela scores 3 is $2 \times \frac{1}{6} = \frac{1}{18}$.
Hence probability required $= \frac{1}{36} \times \frac{1}{18} = \frac{1}{648}$.
This problem could have been solved either using a sample space diagram, or tree diagram. However, the solution here illustrates a method of listing possibilities.

11.6   Statistical Diagrams

It is assumed that the reader has a basic knowledge of how to represent numerical data by means of a diagram. The main types used are bar charts, pie charts, pictograms and histograms. The first three will be considered briefly in this section.

Worked Example 11.4

A survey was carried out of the morning newspapers read by the inhabitants of Steyning. The results were as follows:

Daily Mail	270	*Sun*	340
Guardian	139	*Daily Express*	240
The Times	61	*Daily Mirror*	200
Daily Telegraph	150	*Morning Star*	40

Illustrate this information, using (a) a bar chart, (b) a pie chart, (c) a pictogram.

Solution

In any statistical diagram, clarity is of the utmost importance.
(a) See Fig. 11.5.

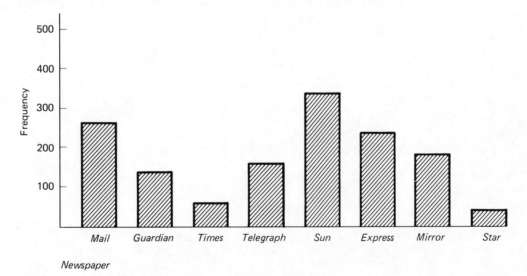

Fig. 11.5

179

(b) When constructing a pie chart, first of all calculate the total number being represented.

$$Total = 270 + 139 + 61 + 150 + 340 + 240 + 200 + 40$$
$$= 1440.$$

This is being represented by $360°$,

hence each degree is $\dfrac{1440}{360} = 4$ people,

or each person is $\frac{1}{4}°$.

The angles are as follows

$$Mail = 270 \times \tfrac{1}{4} = 67.5°,$$

and similarly,

$$Guardian = 34.75°, \quad Times = 15.25°, \quad Telegraph = 37.5°, \quad Sun = 85°,$$
$$Express = 60°, \quad Mirror = 50°, \quad Star = 10°.$$

The pie chart can now be drawn (Fig. 11.6). If the numbers are not drawn on the pie chart, this is a very difficult diagram to read.

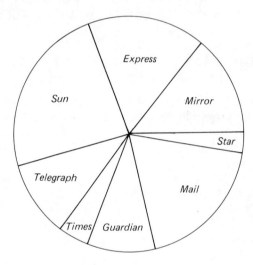

Fig. 11.6

(c) When drawing a pictogram, first choose a motif to represent a certain unit. In this case, we will use 👤 to represent 50 people. The pictogram can now be drawn as in Fig. 11.7.

Mail	👤 👤 👤 👤 👤 👤
Guardian	👤 👤 👤
Times	👤 ∘
Telegraph	👤 👤 👤
Sun	👤 👤 👤 👤 👤 👤 👤
Express	👤 👤 👤 👤 👤
Mirror	👤 👤 👤 👤
Star	👤

👤 represents 50 readers

Fig. 11.7

Accuracy again is limited, although it is better than the pie chart to read.

11.7 Averages

Three types of average are commonly used in everyday life (there are others). The *arithmetic mean* (or mean) of n numbers $a_1, a_2, a_3, \ldots, a_n$ is found by

$$\text{Arithmetic mean} = \frac{a_1 + a_2 + a_3 + \ldots + a_n}{n}.$$

The *mode* of a set of numbers is the number (or numbers) which occurs most often.

The *median* is the middle number (or average of the two middle numbers) if the set of numbers is arranged in increasing order of size.

11.8 Frequency Tables

Numerical data is often much more manageable if it is grouped together in some way. There are two types of situation, as follows:

(a) The marks obtained out of 10 in a single multiple-choice test by 20 students were 8, 6, 5, 0, 1, 4, 3, 4, 6, 7, 6, 5, 5, 6, 3, 4, 7, 6, 9, 10. Calculate (i) the mean score; (ii) the median score; and (iii) the mode, by using a frequency table, as in Table 11.1.

Table 11.1

Score	Tally	Frequency	Score × frequency
0	1	1	0
1	1	1	1
2		0	0
3	11	2	6
4	111	3	12
5	111*	3	15
6	⊥⊥⊤*	5	30
7	11	2	14
8	1	1	8
9	1	1	9
10	1	1	10
	Total	20	105

(i) The mean is found by dividing the sum of score × frequency by the total number of students, not by 10:

$$\text{Mean} = \frac{105}{20} = 5.25.$$

(ii) The median will be the average of the 10th and 11th numbers shown by * in the diagram:

$$\text{Median} = \frac{5 + 6}{2} = 5.5.$$

(iii) The mode is clearly 6.

(b) The heights of a group of 30 children, measured to the nearest centimeter, are shown in Table 11.2.

181

Table 11.2

Height (cm)	Frequency	Cumulative frequency	Middle value	Frequency × middle value
90–99	2	2	94.5	189
100–109	5	7	104.5	522.5
110–119	7	14	114.5	801.5
120–129	10	24	124.5	1245.0
130–139	5	29	134.5	672.5
140–149	1	30	144.5	144.5
Total	30			3575

Calculate (i) the mean height; (ii) the modal height.

(i) The middle value column is obtained from the average of the extremes of the group,

$$\text{i.e. } 94.5 = (90 + 99) \div 2.$$

$$\text{The average height} = \frac{\text{Sum of frequency} \times \text{middle value}}{\text{Total number of children}}$$

$$= \frac{3575}{30} = 119.2 \text{ cm.}$$

(ii) The modal class is clearly 120–129 cm.

11.9 Cumulative Frequency Curves (Ogives)

Referring again to Table 11.2, the third column gives the cumulative frequencies (or running totals). These could be set out as here:

Height (cm)	< 100	< 110	< 120	< 130	< 140	< 150
Cumulative frequency	2	7	14	24	29	30

When plotting the graph of these figures, since the values are given to the nearest cm, the points are plotted at 99.5 cm, 109.5 cm etc.

The cumulative frequency curve (or Ogive) can then be plotted; see Fig. 11.8. **It always has this characteristic shape.**

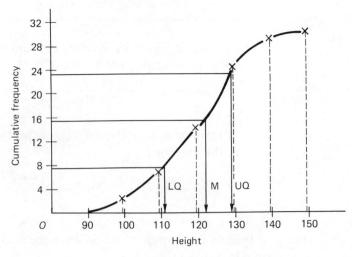

Fig. 11.8

11.10 Information from the Cumulative Frequency Curve

The main types of information that can be gained from these curves are percentiles, median and interquartile range (or semi-interquartile range). If we divide the total frequency into quarters, the value at the lower quarter is referred to as the lower quartile (LQ), the value at the middle gives the median (M), and the value at the upper quarter is the upper quartile (UQ). The *percentiles* are found by dividing the total cumulative frequency into hundredths.

$$\text{The semi-interquartile range (SIQR)} = \frac{\text{Upper quartile} - \text{Lower quartile}}{2}.$$

Worked Example 11.5

Using the information in Fig. 11.8, find (a) the median height, (b) the semi-interquartile range.

Solution

(a) The total frequency is 30.
    The median is between the 15th and 16th values.
    Reading the frequency of 15.5 gives: median height = 122 cm.
(b) The lower quartile is approximately 7.75 (not 7.5).
    The upper quartile is approximately at 23.25 (not 22.5).
    $\therefore$ LQ = 111,
       UQ = 129.

$$\text{SIQR} = \frac{129 - 111}{2} = 9 \text{ cm.}$$

11.11 Assumed Mean

It is sometimes much easier to work out the mean of a frequency table by assuming the mean, and working from this value.

Worked Example 11.6

Calculate the mean of the frequency table (Table 11.1) by assuming the mean is 5.

Solution

Two extra columns are required, one with score minus assumed mean, i.e. $x - A$, the other with $f(x - A)$. See Table 11.3.

Table 11.3

Score x	$x - a$	f	$f(x - A)$
0	−5	1	−5
1	−4	1	−4
2	−3	0	0
3	−2	2	−4
4	−1	3	−3
5	0	3	0
6	1	5	5
7	2	2	4
8	3	1	3
9	4	1	4
10	5	1	5
	Total	20	5
			5

$$\text{Find: } \frac{\text{Total } f(x - A)}{\text{Total f}} = \frac{5}{20} = 0.25.$$

Since we assumed the mean was 5, the correct mean is $A + 0.25 = 5.25$.

11.12 Histograms

A histogram is the display of data in the form of a block graph where the area of each rectangle is proportional to the frequency. When the rectangles are the same width, their heights too are proportional to the frequency and the histogram is synonymous with the bar chart.

An example of a histogram with unequal class intervals follows.

Worked Example 11.7

The following table gives the distribution of the number of employees in the 50 factories in Stenworth:

Number of employees	0–39	40–59	60–79	80–99	100–139
Number of factories	5	15	13	10	7

Construct a histogram to show the distribution.

Solution

The widths of the five class intervals are

$$40 \quad 20 \quad 20 \quad 20 \quad 40$$

Referring to Fig. 11.9, we see that as the widths of the first and last intervals are twice the width of the other three, the heights of the rectangles are reduced in proportion: i.e. if the height of the second rectangle is 15 units, the height of the first rectangle is $\frac{5}{2} = 2\frac{1}{2}$ units and the height of the last rectangle is $\frac{7}{2} = 3\frac{1}{2}$ units.

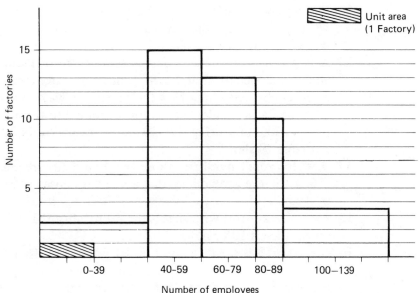

Fig. 11.9

Exercise 11

1 A bag contains 8 red discs and 4 blue discs. If a disc is removed from the bag and not replaced, and then a second disc is removed from the bag, the probability that the two discs are the same colour is:
   A $\frac{2}{3}$ B $\frac{17}{33}$ C $\frac{1}{2}$ D $\frac{5}{9}$

2 The probability of it raining on any day is $\frac{1}{3}$. The probability that it does not rain two days running is:
   A $\frac{2}{3}$ B $\frac{5}{6}$ C $\frac{4}{9}$ D $\frac{1}{2}$

3 The mean of 3 numbers is 14, the mean of 7 other numbers is 20. The mean of the 10 numbers is:
   A 17   B 18.2   C 18   D 16

4 The probability of drawing a picture card from a normal pack of cards is:
   A $\frac{1}{4}$ B $\frac{3}{52}$ C $\frac{3}{13}$ D $\frac{1}{10}$

5 Three coins are tossed together; the probability of obtaining 3 heads is:
   A $\frac{1}{4}$ B $\frac{1}{2}$ C $\frac{1}{6}$ D $\frac{1}{8}$

6 The number which needs to be removed from {11, 15, 2, 18, 14, 3, 6} so that the mean of the remainder is 9 is:
   A 14   B 15   C 6   D none of these

7 The average height of 6 boys is 1.8 m and the average height of 9 girls is 1.5 m. The average height of the group is:
   A 1.65 m   B 1.62 m   C 1.6 m   D 1.7 m

8 The mean of 4, 5, a, b is 10. The mean of 7, 8, a, b is:
   A 11.5   B 12   C 6   D indeterminate

9 The probability that a letter drawn at random from the word *statistics* is a consonant, is:
   A $\frac{3}{10}$ B $\frac{2}{5}$ C $\frac{7}{10}$ D $\frac{1}{2}$

10 A die is rolled twice, and on each occasion it shows 6. If it is rolled a third time, the probability of obtaining a 6 is:
   A $\dfrac{1}{6^3}$ B $\dfrac{1}{6^2}$ C $\frac{1}{6}$ D 0

11 Raffle tickets numbered 1 to 100 are placed in a drum and one is drawn at random. What is the probability that the number is divisible by (a) 2; (b) 5; (c) 2 and 5?

12 The number which needs to be added to the list of numbers $\{1, 8, 15, 9, 11, 18, 15\}$ to increase their average from 11 to 13 is:

   A 12   **B** 11   **C** 27   **D** 13

13 A target at a rifle range consists of a centre of radius 3 cm which scores 100 and two concentric circles of radius 5 cm and 6 cm, which score 50 and 20 respectively. Assuming that the target is always hit, and that the probability of hitting a particular score is proportional to the area for that score, find the probabilities of scoring 100, 50 and 20 respectively. What is the expected total score for 10 shots?

14 $A = \{$four-digit numbers whose first two digits are greater than or equal to $x\}$.

   What is the probability that a number chosen at random from A is divisible by 10?

15 Two dice are thrown; what is the probability that the product of the numbers on the dice is 10 or more?

16 Given that 9 is the mean of 2, x, 10, 12 and 15, find x.

17 If a letter is chosen at random from the word *method*, what is the probability it does not belong to the word *mathematical*?

18 Six equal balls, numbered 1, 2, 3, 4, 5 and 6 respectively, are placed in a bag. A ball is drawn out of the bag at random and then a second ball is drawn out without replacing the first. Find the probability that:

   (a) the first ball is numbered 4;

   (b) the first ball is not the 6;

   (c) the first ball is numbered 2 and the second 5.

19 A bag contains 40 discs: 8 red, 15 green, the rest black.

   (a) If three are drawn out in succession and not replaced, what is the probability of drawing, in order, 2 reds and 1 green?

   (b) If each disc is replaced after drawing, what would the result be?

20 In a class of 30 children, 2 have the first name John and 5 have the surname Jones. Claculate the probability that a child chosen at random does not have the name John Jones. Calculate also the probability that each of the children with the name John has the name John Jones.

   If, further, 2 children have the surname Coates, calculate the probability that a child chosen at random has either the name John Coates or the name John Jones.

21 Four cards numbered 1 to 4 are placed face downwards on a table.

   (a) Two cards are picked up at the same time. Find the number of ways that this can be done. Hence find the probability that if two cards are picked at random from the table, one of them is numbered 2.

   (b) In a second experiment, one card is picked up from the table and marked on the face. The card is replaced, and a second card is picked up and marked. Find the probability that (i) the same card is marked twice; (ii) that the numbers on the two marked cards add up to 5.

22 A target consists of a central circular red area of radius 5 cm which scores 50 marks surrounded by a larger concentric blue circle of radius 20 cm which scores 10 marks. Jane fires a number of shots at the target. Find (a) the probability of hitting the blue area with any shot; (b) the average score per shot she should expect after a large number of attempts.

23 An engineering firm has 60 employees. The histogram, Fig. 11.10, shows the distribution of their salaries (none of which is a whole number of thousands of pounds). The 14 employees earning less than £4000 p.a. work part-time.

   (a) How many of the part-time workers earn more than £1000 p.a.?

   (b) How many employees earn between £4000 and £5000 p.a.?

   (c) How many employees earn between £6000 and £8000 p.a.?

   (d) The 46 full-time workers have a mean salary of £7000 p.a. and the 14

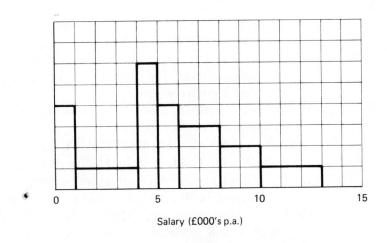

Salary (£000's p.a.)

Fig. 11.10

part-time workers have a mean salary of £1000 p.a. What is the mean salary of the 60 employees? [NISEC]

24 When John and Henry play darts, the probability that John wins a game is 0.7 and that Henry wins a game is 0.3. They play a match to be decided by the first player to win 3 games (the match is finished as soon as this happens). Calculate the probabilities that:
(a) John wins the first three games;
(b) John loses the first game but wins the next three;
(c) John wins by three games to one;
(d) John wins the match in not more than four games. [OLE]

25 The lifetimes in hours of 100 electric light bulbs are decided by the following table:

Lifetimes x	$300 < x \leqslant 400$	$400 < x \leqslant 450$	$450 < x \leqslant 500$	$500 < x \leqslant 550$
Frequency	3	2	6	35

Lifetimes x	$550 < x \leqslant 600$	$600 < x \leqslant 650$	$650 < x \leqslant 700$	$700 < x \leqslant 750$
Frequency	17	10	3	11

Lifetimes x	$750 < x \leqslant 850$	$850 < x \leqslant 1000$
Frequency	10	3

Draw a histogram to illustrate this information.

Estimate the probability that an electric light bulb drawn at random from this sample has a lifetime of over 700 hours.

Calculate an estimate of the mean lifetime of these electric light bulbs.
[OLE]

26 Anne and Jane play a game against each other which starts with Anne aiming to throw a bean bag into a circle marked on the ground.
(a) The probability that the bean bag lands entirely inside the circle is $\frac{1}{2}$, and the probability that it lands on the rim of the circle is $\frac{1}{3}$. Show that the probability that the bag lands entirely outside the circle is $\frac{1}{6}$. What are the probabilities that two successive throws land:
  (i) both outside the circle;
  (ii) the first on the rim of the circle and the second inside the circle?
(b) Jane then shoots at a target on which she can score 10, 5 or 0. With any one shot, the probability that she scores 10 is $\frac{2}{5}$, the probability that she scores 5 is $\frac{1}{10}$, and the probability that she scores 0 is $\frac{1}{2}$. With exactly two shots, what are the probabilities that she scores
(i) 20; (ii) 10?

187

(c) When the bean bag thrown by Anne lands outside the circle, Jane is allowed two shots at her target. If, however, the bean bag lands on the rim of the circle, Jane has one shot, and if it lands inside the circle Jane is not allowed any shots. Find the probability that Jane scores 10 as a result of any one throw from Anne. [O & C]

27 In last year's mathematics examination, there were 20 000 candidates who were awarded grades as follows:

Grade A 2780, Grade B 4360, Grade C 6320,
Grade D 1600, Grade E 1380, Unclassified 3560.

(a) On graph paper, draw a bar chart to represent this information. Use 2 cm to represent 1000 candidates, and 1 cm to represent each grade.

(b) If the information were represented on a pie chart, calculate the angle which would be used for the sector representing Grade C, giving your answer to the nearest degree.

(c) Find, as a decimal correct to three decimal places, the probability that a candidate picked at random will have
(i) Grade C   (ii) Grade A, B or C.

(d) If two candidates are to be picked at random, write down, but do not simplify, an expression for the probability that they will both have Grade C. [L]

28 The matrix shown, denoted by M, is one whose entries are all probabilities relating to chances in the weather:

$$\begin{array}{cc} & \text{Tomorrow} \\ & \begin{array}{cc} \text{Fine} & \text{Wet} \end{array} \\ \text{Today}\quad \begin{array}{c} \text{Fine} \\ \text{Wet} \end{array} & \begin{pmatrix} \frac{2}{3} & \frac{1}{3} \\ \frac{1}{2} & \frac{1}{2} \end{pmatrix} \end{array}$$

For example, the entry in the top left-hand corner indicates that if today's weather is fine, then the probability of tomorrow's being fine is $\frac{2}{3}$.

(a) Write down the probability that if today is fine, tomorrow will be wet.

(b) Assume that today is Sunday and that it is fine. Calculate the probabilities that:
(i) Monday and Tuesday will both be fine;
(ii) Monday will be wet and Tuesday will be fine.

(c) Calculate M^2. Interpret the entry in the bottom right-hand corner of this matrix as a probability.

(d) It is desired to find numbers x and y such that $x + y = 1$ and such that:

$$(x \quad y)\begin{pmatrix} \frac{2}{3} & \frac{1}{3} \\ \frac{1}{2} & \frac{1}{2} \end{pmatrix} = (x \quad y).$$

Verify that $x = \frac{3}{5}$ and $y = \frac{2}{5}$. [OLE]

29 As a test of general knowledge, 200 pupils from a city school had to mark the names of as many streets as they could on a map of the area. The results are given in the following table. For example, 12 pupils named 46 streets correctly, and so on.

Number of streets correct (x)	46	47	48	49	50	51	52	53
Number of pupils (f)	12	25	46	44	30	17	16	10

(a) Draw the frequency polygon of this distribution, using the following scales. On the horizontal axis take values of x from 46 to 53 and a scale of 2 cm to represent 1 street. On the vertical axis take values of f from 0 to 50 and a scale of 2 cm to represent 10 pupils.

(b) For this distribution, find (i) the mode, (ii) the median.

(c) Copy and complete Table 11.4, which uses an assumed mean of 50.

Table 11.4

No. of streets correct x	No. of pupils f	$x - 50$	$f(x - 50)$
46	12	−4	−48
47	25		
48	46		
49	44		
50	30		
51	17		
52	16		
53	10		
Total =	200	Total =	

(d) Hence, or otherwise, calculate the mean of the distribution.   [UCLES]

12 Trigonometry

12.1   Trigonometrical Ratios (Sine, Cosine, Tangent)

There are a number of ways of defining the trigonometrical ratios. The method of coordinates is used here, which also allows extension to any angle.

In Fig. 12.1 *OP* is a line of length *r*, which can rotate in an anticlockwise direction (positive angle).

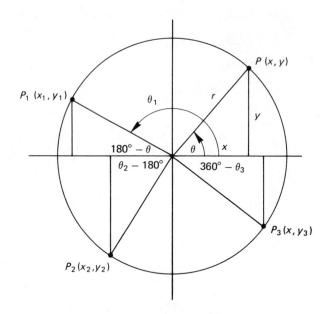

Fig. 12.1

The quadrants of the circle are defined as follows:

$0° \leqslant \theta \leqslant 90°$ is called the *first quadrant*.
$90° < \theta \leqslant 180°$ is called the *second quadrant*.
$180° < \theta \leqslant 270°$ is called the *third quadrant*.
$270° < \theta \leqslant 360°$ is called the *fourth quadrant*.

We have the following definitions:

$$\sin \theta = \frac{y}{r}, \tag{1}$$

$$\cos \theta = \frac{x}{r}, \tag{2}$$

$$\tan \theta = \frac{y}{x}. \tag{3}$$

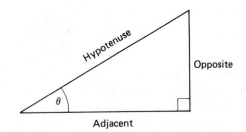

Fig. 12.2

If we restrict ourselves to the first quadrant, we find as in Fig. 12.2:

$$\sin \theta = \frac{\text{opposite}}{\text{hypotenuse}}, \qquad (1A)$$

$$\cos \theta = \frac{\text{adjacent}}{\text{hypotenuse}}, \qquad (2A)$$

$$\tan \theta = \frac{\text{opposite}}{\text{adjacent}}. \qquad (3A)$$

Formulae (1)–(3) can be rewritten:

$$y = r \sin \theta, \quad (4) \qquad x = r \cos \theta, \quad (5) \qquad y = x \tan \theta. \quad (6)$$

or $\qquad r = \dfrac{y}{\sin \theta}, \quad (7) \qquad r = \dfrac{x}{\cos \theta}, \quad (8) \qquad x = \dfrac{y}{\tan \theta}, \quad (9).$

Note: In more advanced work:

$$\frac{1}{\sin \theta} = \operatorname{cosec} \theta, \quad (10) \qquad \frac{1}{\cos \theta} = \sec \theta, \quad (11) \qquad \frac{1}{\tan \theta} = \cot \theta. \quad (12)$$

hence $\quad r = y \operatorname{cosec} \theta, \quad (13) \qquad r = x \sec \theta, \quad (14) \qquad x = y \cot \theta, \quad (15).$

In the second quadrant, when P reaches P_1 after rotating an angle θ_1 then y_1 is positive but x_1 is negative.

$$\sin \theta_1 = \frac{y_1}{r} \quad \text{is still positive,}$$

$$\cos \theta_1 = \frac{x_1}{r} \quad \text{is negative,}$$

and $\qquad\qquad \tan \theta_1 = \dfrac{y_1}{x_1} \quad \text{is negative.}$

But $\dfrac{y_1}{r} = \sin (180 - \theta_1), \quad \therefore \sin \theta_1 = \sin (180 - \theta_1),$ $\qquad\qquad (16)$

$\dfrac{-x}{r} = \cos (180 - \theta_1), \qquad \therefore \cos \theta_1 = -\cos (180 - \theta_1),$ $\qquad (17)$

$\dfrac{-y}{x} = \tan \theta_1, \qquad\qquad \therefore \tan \theta_1 = -\tan (180 - \theta_1).$ $\qquad (18)$

In the third quadrant, when P reaches P_2 after rotating an angle θ_2, then y_2 is negative and x_2 is negative.

We find $\qquad\qquad \cos \theta_2 = -\cos (\theta_2 - 180),$ $\qquad\qquad (19)$
$$\sin \theta_2 = -\sin (\theta_2 - 180), \qquad\qquad (20)$$
$$\tan \theta_2 = \tan (\theta_2 - 180). \qquad\qquad (21)$$

In the fourth quadrant, when P reaches P_3 after rotating an angle θ_3, then y_3 is negative and x_3 is positive.

We find that
$$\cos\theta_3 = \cos(360 - \theta_3), \tag{22}$$
$$\sin\theta_3 = -\sin(360 - \theta_3), \tag{23}$$
$$\tan\theta_3 = -\tan(360 - \theta_3). \tag{24}$$

We illustrate the use of these formulae by the following examples:
(a) $\cos 127° = -\cos(180° - 127°) = -\cos 53° = -0.6018$ (see equation 17).
(b) $\tan 210° = \tan(210° - 180°) = \tan 30° = 0.5774$ (see equation 21).
(c) $\sin 290° = -\sin(360° - 290°) = -\sin 70° = -0.9397$ (see equation 23).

The final answers can be obtained directly by using the calculator

The reverse process is slightly more involved.
Find x if:
(a) $\cos x = -0.5$,   (b) $\sin x = 0.8$,   (c) $\tan x = 2.3$.
(a) Since $\cos x$ is negative, x could be in the second or third quadrant. On the calculator, first find the inverse cosine of 0.5; this gives 60°.

Hence: $x = 180° - 60° = 120°$,
or $x = 180° + 60° = 240°$.

(b) Since $\sin x$ is positive, x could be in the first or second quadrant. On the calculator, first find the inverse sine of 0.8; this gives 53.1°.

hence: $x = 53.1°$,
or $x = 180° - 53.1° = 126.9°$.

(c) Since $\tan x$ is negative, x could be in the second or fourth quadrant. First find the inverse tangent of 2.3; this gives 66.5°.

hence: $x = 180° - 66.5° = 113.5°$
or $x = 360° - 66.5° = 293.5°$.

The graphs of $\sin x$, $\cos x$ and $\tan x$ are shown in Fig. 12.3 for $0 \leqslant x \leqslant 360°$. The functions can be extended in either direction.

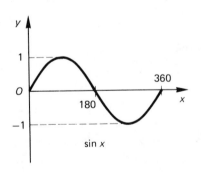
sin x

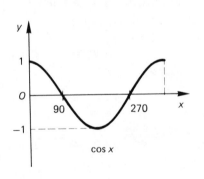
cos x

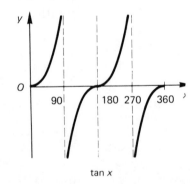
tan x

Fig. 12.3

12.2   Pythagoras' Theorem

In any right-angled triangle ABC ($\angle ABC = 90°$), Pythagoras' theorem states that:
$$AC^2 = AB^2 + BC^2. \tag{25}$$

The proof given here relies on ideas developed in Chapter 9. ABC is a right-angled triangle (Fig. 12.4). Consider squares drawn on the sides AB, BC and CA.

Proof:
Draw XX' at right angles to AC. The area of $ABB'A'$ = area $ABXR$ (shear parallel to AB).

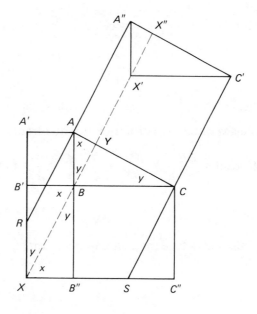

Fig. 12.4

The area of $BCC''B''$ = area $BCSX$ (shear parallel to CB).
$ABC \to BB'X$ (by a rotation of 90°),

$$\therefore XB = CC'.$$

$ABC \to A''X'C'$ (by a translation $\overrightarrow{CC'}$),

$$\therefore \text{area } BCSX = \text{area } BCC'X' = \text{area } YCC'X'' \text{ (shear parallel to } CC').$$

Similarly,

$$\text{area } ABX'A'' = \text{area } ABXR = \text{area } AYX''A'' \text{ (shear parallel to } AA'),$$
$$\therefore \text{area } BCC''B'' + \text{area } ABB'A' = \text{area } AA''C'C.$$

$$\therefore AC^2 = AB^2 + BC^2.$$

12.3   Relationship Between sin x, cos x and tan x

Pythagoras' theorem applied to the triangle in Fig. 12.5 gives:

$$c^2 = a^2 + b^2. \tag{26}$$

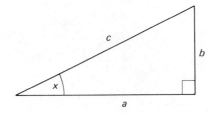

Fig. 12.5

Divide equation (26) by c^2:

$$1 = \frac{a^2}{c^2} + \frac{b^2}{c^2}.$$

But
$$\frac{a}{c} = \cos x, \therefore \frac{a^2}{c^2} = (\cos x)^2, \text{ written } \cos^2 x,$$

$$\frac{b}{c} = \sin x \therefore \frac{b^2}{c^2} = (\sin x)^2, \text{ written } \sin^2 x,$$

hence
$$1 = \cos^2 x + \sin^2 x. \tag{27}$$

This is in fact true for x in any quadrant.

Consider $\dfrac{\sin x}{\cos x} = \dfrac{b/c}{a/c} = \dfrac{b}{\not{c}} \times \dfrac{\not{c}}{a} = \dfrac{b}{a} = \tan x;$

$$\therefore \tan x = \frac{\sin x}{\cos x}. \tag{28}$$

This is also true for x in any quadrant.

Worked Example 12.1

If $\cos x = \dfrac{1}{\sqrt{5}}$, and $x \geqslant 180°$, find

(a) $\sin x$;   (b) $\tan x$.

In any question of this type, make sure that you determine the quadrants that any angle must be in.

Solution

(a) Since $\cos x$ is positive and $x \geqslant 180°$, x must be in the *fourth* quadrant, hence $\sin x$ is negative.

$$1 = \cos^2 x + \sin^2 x = \left(\frac{1}{\sqrt{5}}\right)^2 + \sin^2 x,$$

$$\therefore 1 - \frac{1}{5} = \sin^2 x,$$

$$\therefore \sin^2 x = \frac{4}{5},$$

$$\therefore \sin x = \frac{-2}{\sqrt{5}}.$$

(b) $\tan x = \dfrac{\sin x}{\cos x} = \dfrac{-2/\sqrt{5}}{1/\sqrt{5}} = -2.$

12.4   Solution of Right-angled Triangles

The ideas of sections 12.1 and 12.2 are now used in the solution of problems as indicated below.

Worked Example 12.2

The diagram in Fig. 12.6 shows part of a girder. Find:
(a) $\angle RPQ$,   (b) RQ,   (c) SQ.

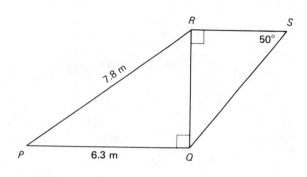

Fig. 12.6

Solution

Since two sides are given in triangle PRQ, either the sine, the cosine or the tangent ratio can be used. Relative to $\angle RPQ$, 7.8 is the hypotenuse and 6.3 is the *adjacent*, hence the cosine ratio is needed.

(a) Using formula (2A), $\cos R\hat{P}Q = \dfrac{PQ}{PR} = \dfrac{6.3}{7.8} = 0.8077$.

Pressing the inverse cosine button on the calculator, we get:

$\angle RPQ = 36.1°$.

(b) The easiest way to find RQ is using Pythagoras' theorem:

$$PR^2 = RQ^2 + PQ^2,$$

$$\therefore \quad 7.8^2 = RQ^2 + 6.3^2,$$

$$\therefore \quad 60.84 = RQ^2 + 39.69$$

$$\therefore \quad 60.84 - 39.69 = 21.15 = RQ^2.$$

Remember, Pythagoras' theorem gives the square of the side. The square root must now be found:

$$RQ = \sqrt{21.15} = 4.599 \text{ m}$$

(c) In triangle RQS, we now have one side and one angle, hence we can use formulae (4)–(9). RQ is *opposite*, and SQ is the hypotenuse. Since the hypotenuse is unknown, we require formula (7).

$$\therefore QS = \frac{QR}{\sin RSQ} = \frac{4.599}{\sin 50°} = \frac{4.599}{0.766},$$

$$\therefore QS = 6.004 \text{ cm}.$$

12.5 Special Triangles

There are certain triangles which occur frequently in examination questions where the sides or angles are related in a particularly simple way. The triangles are often disguised (see Worked Example 12.3). Figure 12.7 shows a number of these triangles.

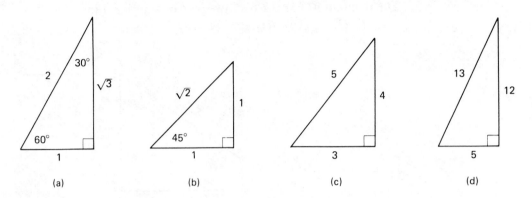

Fig. 12.7

Worked Example 12.3

In Fig. 12.8, $ABCD$ is a square of side 24 cm. $\angle EAC = 90°$. Calculate EC.

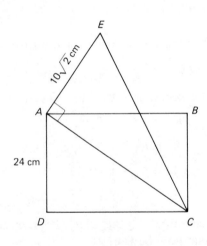

Fig. 12.8

Solution

The fact that $\sqrt{2}$ appears in the diagram immediately suggests one of the special triangles.

In triangle EAC only one side is known. Hence, before EC can be found, we must find AC. $\triangle ADC$ is an enlargement, scale factor 24, of triangle (b) in Fig. 12.7.

$$\therefore AC = 24\sqrt{2} \text{ cm.}$$

Careful observation of $\triangle AEC$ shows that it is an enlargement of triangle (d) Fig. 12.7 with scale factor $2\sqrt{2}$.

Hence $\qquad\qquad\qquad\qquad EC = 26\sqrt{2} \text{ cm.}$

12.6   Angles of Elevation and Depression

Figure 12.9 illustrates the meaning of the two terms

angle of elevation (measured upward from the horizontal),

and   *angle of depression* (measured downward from the horizontal).

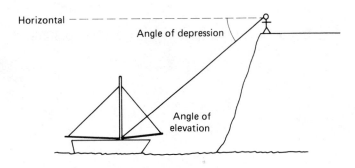

Fig. 12.9

It can be seen that:

$$\frac{\text{the angle of depression}}{\text{of the boat from the man}} = \frac{\text{the angle of elevation}}{\text{of the man from the boat.}}$$

12.7   Bearings

Referring to Fig. 12.10, the bearing of *B* **from** *A* (always measured clockwise) is 025° (three figures are always given). To find the bearing of *A* **from** *B*, we need to find *y*.
  Clearly, $y = 180° + 25° = 205°$.

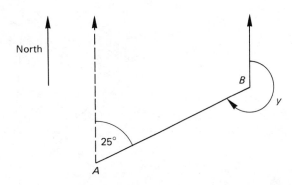

Fig. 12.10

Note: Bearings are still sometimes given using the compass points. However, they are not used in this book.

Worked Example 12.4

The points *B*, *P* and *Q* lie in the same horizontal plane, and *AB* is a vertical monument of height 75 m. From the point *A* at the top of the monument, the angles of

197

depression of the points P and Q are $17°$ and $20°$ respectively. Calculate, to the nearest metre, the distances BP and BQ.

The bearings of P and Q from B are $026°$ and $116°$ respectively. Calculate the distance PQ to the nearest metre, and the bearing of Q from P to the nearest degree. [London]

Solution

In Fig. 12.11, we have $\dfrac{75}{PB} = \tan 17° = 0.3058$.

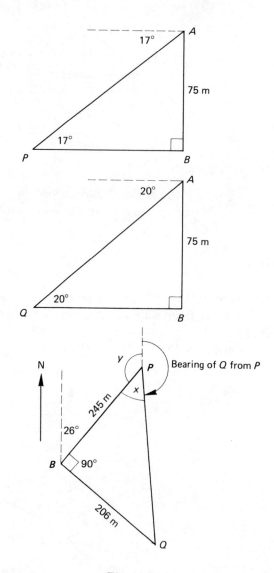

Fig. 12.11

$\therefore PB = \dfrac{75}{0.3058} = 245$ m.

Also, $\dfrac{75}{QB} = \tan 20° = 0.3640$,

$\therefore QB = \dfrac{75}{0.3640} = 206$ m.

Note: $\angle PBQ = 90°$ because the bearing of Q from B is $116°$ and $26° + 90° = 116°$.

Using Pythagoras,

$$PQ^2 = 245^2 + 206^2 = 102\,461,$$

$$\therefore \quad PQ = \sqrt{102\,461} = 320 \text{ m}.$$

To find the bearing of Q from P, we must first find x.

$$\tan x = \frac{206}{245} = 0.8408,$$

$$\therefore \quad x = 40° \text{ (nearest degree)}.$$

We know $y + 26 = 180$, $\therefore y = 154$,

hence the bearing of Q from P is $360° - (154° + 40°) = 166°$.

12.8   Reduction to Right-angled Triangles

There are many situations where triangles which at first sight are not right-angled triangles can be reduced to right-angled triangles. The following examples should illustrate this.

Worked Example 12.5

The diagram in Fig. 12.12 shows the cross-section of a cylindrical oil-storage tank of radius 2.6 m. Find the depth of the oil above the bottom of the tank when $\angle AOB = 109°$.

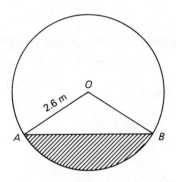

Fig. 12.12

Solution

First recognize that triangle AOB is an isosceles triangle. If divided in half, it becomes two right-angled triangles, as in Fig. 12.13.

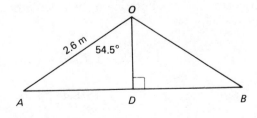

Fig. 12.13

199

$$\cos 54.5° = \frac{OD}{2.6}, \therefore OD = 2.6 \times 0.5807$$

$$= 1.509.$$

To get the depth of the oil, we must subtract OD from the radius:

$$\text{depth of oil} = 2.6 - 1.509 = 1.091 \text{ m.}$$

The next example is slightly more complicated.

Worked Example 12.6

In order to measure the height of a vertical tower OT which is in an inaccessible position, the angles of elevation of the top of the tower are taken from two points A and B distance 100 m apart as shown in Fig. 12.14.
Find the height of the tower.

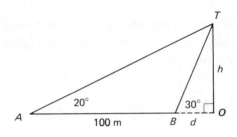

Fig. 12.14

Solution

This question can be solved using the sine rule (see section 12.10). The following method involves simultaneous equations.

In triangle BTO, $\qquad\qquad \frac{h}{d} = \tan 30° = 0.5774,$

$$\therefore \quad \frac{h}{0.5774} = d. \qquad\qquad (1)$$

In triangle TAO, $\qquad\qquad \frac{h}{(d+100)} = \tan 20° = 0.3640,$

$$\therefore \quad \frac{h}{0.364} = d + 100. \qquad\qquad (2)$$

Subtract (1) from (2):

$$\frac{h}{0.364} - \frac{h}{0.5774} = 100.$$

By using the $1/x$ button on the calculator or reciprocal tables, this can more easily be written:

$$2.747h - 1.732h = 100, \qquad \left[\frac{1}{0.364} = 2.747 \text{ etc.}\right]$$

$$\therefore 1.015h = 100,$$

$$\therefore h = \frac{100}{1.015} = 98.52 \text{ m.}$$

12.9   Perpendicular Height of a Triangle

The following method is a useful way of finding the perpendicular height of a triangle, given the three sides of the triangle. It will be illustrated by a worked example.

Worked Example 12.7

In triangle ABC, Fig. 12.15, $AB = 8.3$ cm, $CB = 4.9$ cm and $AC = 6.7$ cm. Find the perpendicular distance of C from AB.

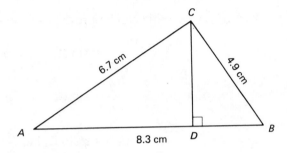

Fig. 12.15

Solution

Referring to the diagram, we are trying to find CD.
We begin by finding the area of the triangle using the formula

$$\text{Area} = \sqrt{s(s-a)(s-b)(s-c)} \text{ (see Chapter 4),}$$

$$s = \tfrac{1}{2}(6.7 + 8.3 + 4.9) = 9.95.$$

$$\text{Area} = \sqrt{9.95 \times 5.05 \times 3.25 \times 1.65} = 16.42 \text{ cm}^2.$$

Using $\tfrac{1}{2}$ base $\times$ height = area, we get $\tfrac{1}{2} \times 8.3 \times CD = 16.42$.

$$\therefore 4.15CD = 16.42, \therefore CD = \frac{16.42}{4.15} = 3.96 \text{ cm.}$$

12.10   The Sine Rule

The first of the rules which can be used in solving non-right-angled triangles is the sine rule.
The rule states:

$$\frac{a}{\sin A} = \frac{b}{\sin B} = \frac{c}{\sin C} = 2R \tag{29}$$

where R is the radius of the circumscribing circle through A, B and C; see Fig. 12.16.

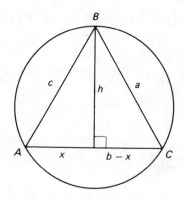

Fig. 12.16

Part of the formula can be proved as follows:

$$\frac{h}{c} = \sin A, \qquad \therefore h = c \sin A\,;$$

$$\frac{h}{a} = \sin C, \qquad \therefore h = a \sin C\,;$$

$$\therefore \ c \sin A = a \sin C, \ \text{i.e.} \ \frac{a}{\sin A} = \frac{c}{\sin C}\,.$$

The sine rule should be used if two angles and one side are given, or two sides and one angle which is opposite either of the two given sides.

12.11   Sine Rule (Ambiguous Case)

When using the sine rule to find angles, it is important to have an approximate idea of the answer, as in some cases more than one solution is possible.

Worked Example 12.8

In triangle XYZ, $\angle XYZ = 30°$, $XY = 8$ cm, and $XZ = 5$ cm. Find the two possible values of XYZ and illustrate with a diagram.

Solution

If you try to draw triangle XYZ, then the two possible triangles can be seen in Fig. 12.17: XYZ_1 and XYZ_2.

$$\frac{XY}{\sin Z} = \frac{XZ}{\sin Y}, \qquad \therefore \ \frac{XY}{XZ} \sin Y = \sin Z\,;$$

$$\sin Z = \frac{8 \sin 30°}{5} = 0.8.$$

Since $\sin Z$ is positive, Z could be in the first or second quadrant.
  Hence, $Z_1 = 53.1°$,
      $Z_2 = 126.9°$.

202

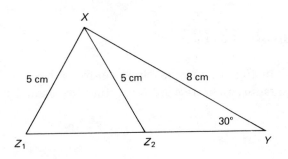

Fig. 12.17

12.12　The Cosine Rule

The cosine formula is used if either three sides of a triangle are given, or two sides and the angle between the two sides.

With the notation of section 12.10:

$$a^2 = b^2 + c^2 = 2bc \cos A$$

or

$$\cos A = \frac{b^2 + c^2 - a^2}{2bc}.$$

Proof:

$$c^2 = h^2 + x^2 \qquad \text{and} \qquad a^2 = h^2 + (b - x)^2 ;$$

$$c^2 - x^2 = a^2 - (b - x)^2 = a^2 - (b^2 + x^2 - 2bx)$$

$$c^2 = a^2 - b^2 + 2bx \qquad \text{but} \qquad x = c \cos A,$$

hence

$$a^2 = c^2 + b^2 - 2bc \cos A.$$

Remember, for an obtuse angle cos A will be negative.

Worked Example 12.9

In triangle PQR, PQ = 5 cm, QR = 6 cm and RP = 7 cm. Find the smallest angle of the triangle.

Solution

The order of size of the angles of a triangle is always the largest angle opposite the largest side and the smallest angle opposite the smallest side.

We need to find the angle opposite PQ, which is R.

$$\cos R = \frac{6^2 + 7^2 - 5^2}{2 \times 6 \times 7} = \frac{60}{84} = 0.7143,$$

$$\therefore \angle R = 44.4°.$$

Worked Example 12.10

In Fig. 12.18, the circle centre A has radius 3 cm, and the circle centre B has radius 5 cm. If $AB = 6$ cm, find the shaded area common to both circles.

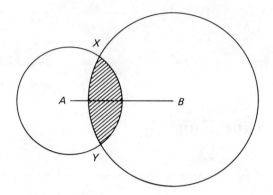

Fig. 12.18

Solution

This problem is more involved than it might seem at first sight. We begin by joining XY and looking at half of the diagram as shown in Fig. 12.19.

The shaded area = Area of sector AXY − Area of triangle AXY.

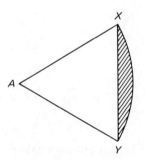

Fig. 12.19

The next problem to overcome is that we need angle XAY, and this cannot be found from triangle AXY as XY is not known. We now go to triangle AXB; see Fig. 12.20.

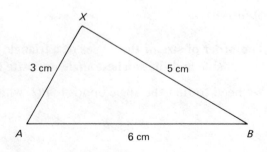

Fig. 12.20

Using the cosine rule for angle A, we have:

$$\cos A = \frac{6^2 + 3^2 - 5^2}{2 \times 6 \times 3} = \frac{20}{36} = 0.5556,$$

hence $\qquad\qquad A = 56.2°.$

$\angle XAY = 2 \times 56.2° = 112.4°$

The shaded area in Fig. 12.19 $= \pi \times 3^2 \times \dfrac{112.4}{360} - \dfrac{1}{2} \times 3^2 \times \sin 112.4°$

$$= 4.67 \text{ cm}^2.$$

A similar method on the other half gives

$$\text{Total area} = 4.67 + 2.24 = 6.91 \text{ cm}^2.$$

12.13 Three-dimensional Problems

In order to tackle most three-dimensional problems at this level, it is necessary to have a clear understanding of how to find the angle between a line and a plane, and also the angle between two planes.

Figure 12.21 shows a line L passing through a plane P, cutting the plane at O. N is a point in the plane, and Q a point on the line, and $\angle QNO = 90°$. QN is *perpendicular to the plane*. For this position, angle QON is the angle between the line and the plane.

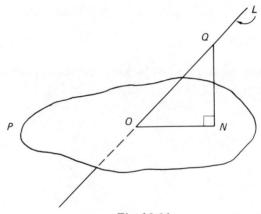

Fig. 12.21

Figure 12.22 shows two planes P_1 and P_2 which have a common line m. q is a line in P_1 at right angles to m, and n is a line in P_2 at right angles to m. The angle between n and q is the angle between the planes.

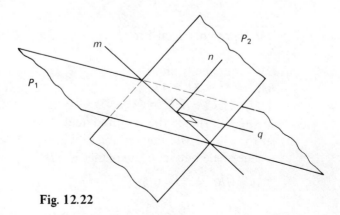

Fig. 12.22

Worked Example 12.11

In Fig. 12.23 $ABCDA'B'C'D'$ is a cube of side 6 cm. M is the midpoint of CC'. Find

 (i) the angle between AM and the plane $BB'C'C$,

(ii) the angle between the planes ABM and $ABCD$.

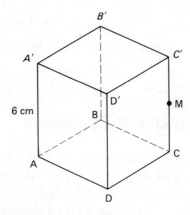

Fig. 12.23

Solution

The simplest way of tackling a question similar to part (i) is to draw the equivalent triangle to QON which appears in Fig. 12.24.

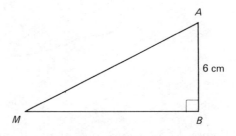

Fig. 12.24

(a) AM represents L in Fig. 12.21 and $BB'C'C$ represents P, hence B is the point N.

BM^2 must be found first:

$BM^2 = 6^2 + 3^2 = 45$.

$AM^2 = AB^2 + BM^2 = 6^2 + 45 = 81$,

$\therefore AM = 9$. Hence the required angle $A\hat{M}B$ can be found.

(b) In Fig. 12.22 the lines n and q will be BM and BC, because AB is the line m.

The angle between the planes is MBC.

$\therefore \tan M\hat{B}C = \dfrac{3}{6} = 0.5$,

$\therefore \angle MBC = 26.6°$.

12.14 Latitude and Longitude

Another problem in three-dimensional work is linked to the earth, considered as a perfect sphere of approximate radius 6370 km. See Fig. 12.25 for definitions.

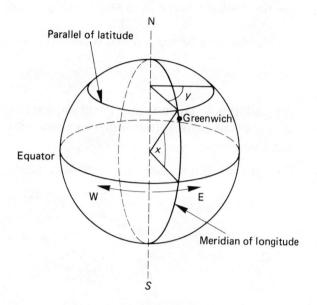

x is the angle of latitude (N or S)

y is the angle of longitude (W or E)

Fig. 12.25

Great circle: Any circle on the surface of the earth whose centre is the centre of the earth, including circles of longitude, and the equator, is called a *great circle*.

Nautical mile: The length of arc on a great circle which subtends an angle of 1′ at the centre is called a *nautical mile* (n mile).

Radius of circle of latitude: In Fig. 12.26, if A is in latitude $x°$ N, r is the radius of the circle of latitude, and R is the radius of the earth, then

$$\frac{r}{R} = \cos x°, \ \therefore \ r = R \cos x°.$$

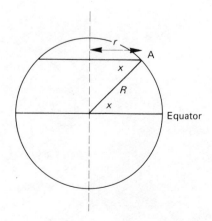

Fig. 12.26

Worked Example 12.12

A is the point (18° E, 60° N), B is the point (12° W, 60° N) and C is the point (12° W, 18° S). Find

(a) the distance between A and B along a circle of latitude, measured in nautical miles,

(b) the distance between B and C along a circle of longitude measured in km.

Solution

All problems in latitude and longitude are best solved by drawing either the relevant circle of latitude (remember its radius may have to be found), or the circle of longitude.

(a) r is not required if we are working in nautical miles. Angle between A and B is

$$12° + 18° = 30° = 30° \times 60 = 1800'.$$

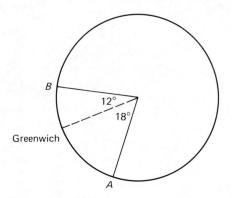

Fig. 12.27

Since we are not on a great circle,

the distance in nautical miles = 1800 cos 60° = 900 n miles.

(b) Angle between B and C is 60° + 18° = 78°;

$$\text{distance } BC = 2 \times \pi \times 6370 \times \frac{78}{360} = 8672 \text{ km.}$$

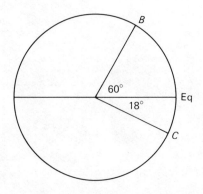

Fig. 12.28

12.15 Velocity Triangles

In Chapter 7, it was shown how to use the method of vectors in constructing velocity triangles. We will now illustrate how the method of trigonometry can be used in the calculation of such triangles, using the following examples.

Worked Example 12.13

A plane sets out to fly from X to Y which is 500 km due east of X. The plane has an airspeed of 350 km/h and there is a wind of 50 km/h blowing in the direction 130°. Find how long it takes.

Solution

The velocity triangle is shown in Fig. 12.29. We need to find the groundspeed PR, before the time can be calculated.

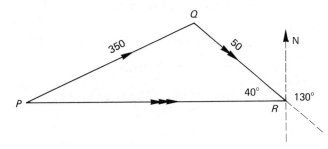

Fig. 12.29

We first find angle P:

$$\frac{50}{\sin P} = \frac{350}{\sin 40°}, \therefore \sin P = \frac{50 \sin 40°}{350}, \text{ hence } P = 5.3°.$$

We can find Q since $Q = 180° - 40° - 5.3° = 134.7°$.
$PR^2 = 50^2 + 350^2 - 2 \times 50 \times 350 \cos 134.7°$,
hence $PR = 386.8$.

$$\text{Time} = \frac{\text{distance}}{\text{speed}} = \frac{500}{386.8} = 1.29 \text{ hours}.$$

Worked Example 12.14

A, B and C are three airfields. B is 200 km due south of A and C is 140 km due west of B. There is an easterly wind of 60 km/h blowing when a light aircraft leaves A in order to fly to B.
(a) If the airspeed of the plane is 100 km/h, find how long it takes to reach B.
(b) If the plane sets a westerly course in order to fly from B to C, but the wind has now changed and is blowing from the north, find the position of the air- craft relative to C after 1 hour.

Always establish whether the course or the track is known, each requires a different approach.

Solution

(a) The track is known, in this case due south. The velocity triangle must look like Fig. 12.30.

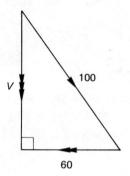

Fig. 12.30

If V is the groundspeed,
$$V^2 = 100^2 - 60^2 = 6400$$
$$\therefore \quad V = 80.$$

The plane is travelling a distance of 200 km due south,
$$\therefore \text{ Time} = \frac{200}{80} = 2\tfrac{1}{2} \text{ hours.}$$

(b) In this part, we know the course; the velocity diagram can be seen in Fig. 12.31.

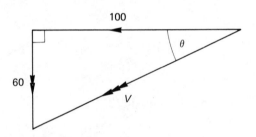

Fig. 12.31

$$V^2 = 60^2 + 100^2 = 13\,600,$$
$$\therefore \quad V = 116.6.$$
$$\tan \theta = \frac{60}{100} = 0.6$$
$$\therefore \theta = 30.96°.$$

The aeroplane flies on a track of $270 - 30.96° = 239.04°$, with a groundspeed of 116.6 km/h. The position D on the ground after 1 hour is shown in Fig. 12.32.

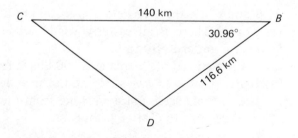

Fig. 12.32

Using the cosine rule, $CD^2 = 140^2 + 116.6^2 - 2 \times 140 \times 116.6 \cos 30.96$,
$\therefore CD = 72.1$

Using the sine rule, $\dfrac{\sin C}{116.6} = \dfrac{\sin 30.96}{72.1}$,

$\therefore \sin C = \dfrac{116.6 \sin 30.96}{72.1}$,

$\therefore C = 56.3°$.

The position of the aeroplane is therefore 72.1 km from C on a bearing of 146.3°.

Exercise 12

1 $ABCD$ is a square of side 2 cm. P, Q, R and S are the mid-points of AB, BC, CD and DA respectively.
$AP + AQ + AR + AS$ equals:
  A   6 cm   B $2 + 2\sqrt{5}$ cm   C $2 + \sqrt{5}$ cm   D $2(1 + 2\sqrt{5})$ cm

2 Given that $PR = 4$ cm, Fig. 12.33, an expression in cm for the length of QS is:

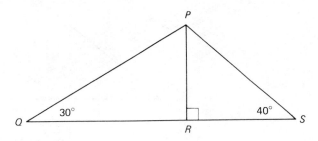

Fig. 12.33

A $\dfrac{4}{\tan 30°} + \dfrac{4}{\tan 40°}$ B $4 \tan 30° + 4 \tan 40°$

C $\dfrac{4}{\sin 30°} + \dfrac{4}{\sin 40°}$ D $\dfrac{4}{\cos 30°} + \dfrac{4}{\cos 40°}$ [AEB]

3 Given that $\tan \alpha = 5/3$, $\cos^2 \alpha$ equals.
  A   9/34   B   9/16   C   25/34   D   25/16   [AEB]

4 In Fig. 12.34, $\tan x$ equals:
  A a/b B $-a/b$ C b/a D $-b/a$

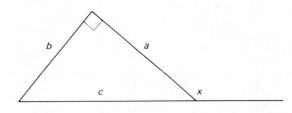

Fig. 12.34

5 The length of the diagonal AB of the rectangular box shown in Fig. 12.35 is:
  A   7 cm   B   5 cm   C $\sqrt{29}$ cm   D   9 cm

211

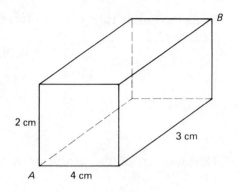

Fig. 12.35

6 In Fig. 12.36, PQ is parallel to SR; PQ = 10 cm, QR = 12 cm, PR = 15 cm. Given that $\angle SPR = \angle PQR$, the length, in cm, of PS is:

    A   8.0   **B**   12.5   **C**   13.5   **D**   18.0   [AEB]

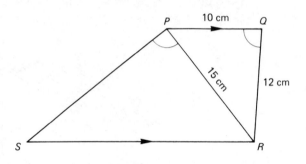

Fig. 12.36

7 Fig. 12.37 shows the position of 3 yachts in a race, A, B, C. The bearing of B from A is 040°, and the bearing of B from C is 310°. It follows that the bearing of A from C must be:

    A   350°   **B**   260°   **C**   270°   **D**   none of these

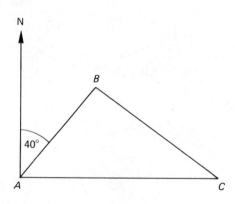

Fig. 12.37

8 In triangle ABC, it was found that $AB^2 < AC^2 + BC^2$. It follows that

    A $\angle C$ is obtuse   **B** $\sin C < \frac{1}{2}$ **C** $\cos C > 0$ **D** $\angle ABC = 90°$

9 A is the point (10° E, 60° N) and B is the point (14° W, 60° N). The shortest distance measured in nautical miles between A and B along a circle of latitude is:

A 240   **B** 1440   **C** 120   **D** 720

10 In Fig. 12.38, ABD is a straight line, and angle x is an acute angle such that $\sin x = \frac{3}{5}$. It follows that AC equals:

A 5 cm   **B** $\sqrt{5.8}$ cm   **C** $\sqrt{44.2}$ cm   **D** 6 cm

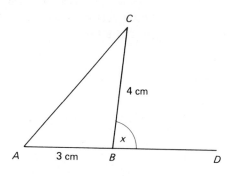

Fig. 12.38

11 In Fig. 12.39, calculate:
(a) the length of PS
(b) the size of $\angle SRP$
(c) the length of RQ

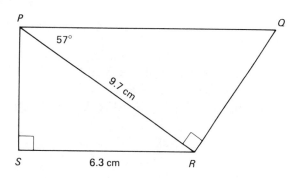

Fig. 12.39

12 In the diagram Fig. 12.40, ABC is a straight line, $BE = 20$ cm, and $BD = 10$ cm. $\angle EAB = \angle BCD = 90°$. Calculate:
(a) AB (b) DC (c) $\tan x$ (d) x [AEB]

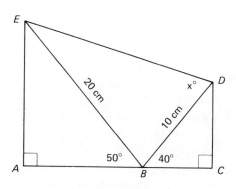

Fig. 12.40

13 In the kite $ABCD$, where $AB = AD$, the diagonals BD and AC meet at E; P is the mid-point of AE and Q is the mid-point of CE. Given that $BD = 12$ cm, $AC = 13$ cm and $AE = 4$ cm:
(a) calculate AB and CB in cm;
(b) prove $\angle ABC = 90°$;
(c) calculate $\angle EBC$;
(d) calculate $\angle PBQ$. [L]

14 In Fig. 12.41 $ABCD$ represents part of the cross-section of a circus tent with BC and AD being vertical poles. $FECD$ is level ground and AF and BE are guy-ropes to support the tent.

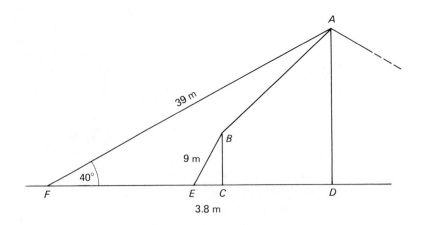

Fig. 12.41

$AF = 39$ m, $BE = 9$ m, $EC = 3.8$ m and $\angle AFE = 40°$. Calculate:
(a) the height of the tent AD;
(b) the size of $\angle BEC$. [WJEC]

15 In Fig. 12.42, $QM = MR$. Calculate QR.

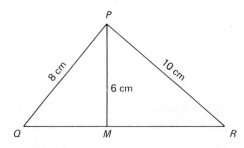

Fig. 12.42

16 If $\sin^2 \alpha = \frac{5}{9}$ and α is obtuse, calculate the value of $\cos \alpha$.

17 (a) Find the obtuse angle whose tangent is -1.5.
(b) If α is an acute angle and $\sin \alpha = 24/25$, calculate $\cos \alpha$.

18 Two ships X and Y leave a port P at 12.00 hours. Ship X sails at 18 knots on a bearing $040°$, and ship Y sails at 20 knots on a constant bearing so that Y is always due south of X. Calculate:
(a) the bearing on which ship Y is sailing;
(b) the distance between the ships at 14.00 hours.
(c) At 14.00 hours, X and Y are 70 nautical miles and 50 nautical miles

respectively from a lighthouse L which is east of the line XY. Calculate the bearing of L from the ship Y.   [L]

19 A, B and C are three points in a straight line on horizontal ground and CD is a vertical rock face, as shown in Fig. 12.43.

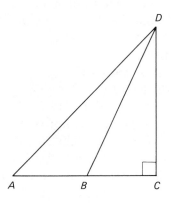

Fig. 12.43

Bruce and Stephanie each wish to estimate the height of the rock face. Their methods are as follows:

(a) Bruce estimates that, at A, the angle of elevation of D is $20°$, the distance AB is 200 m and the distance BC is 500 m. Using Bruce's estimates, calculate:
 (i) the height of the rock face;
 (ii) the angle CBD.

(b) Stephanie estimates that, at A, the angle of elevation of D is $20°$, the distance AB is 200 m and angle CBD is $25°$. Let $BC = d$ metres and $CD = h$ metres. Using Stephanie's estimates, write down two equations involving d and h. Hence calculate the values of d and h.   [OLE]

20 A, B and C are three points on a map. $AB = 6.8$ cm, $AC = 5.3$ cm and $\angle BAC = 44.6°$.

(a) Calculate the length of BC.

(b) Given that the actual distance represented by AB is 13.6 km, calculate the scale of the map in the form 1 : n, and the actual distance represented by BC in kilometres.

(c) Given that B is due east of A and that C is north of the line AB, calculate the bearing of C from A.

21 (a) Find two angles whose sine is 0.55.

(b) Figure 12.44, which is not drawn to scale, shows two points A and B on a coastline which runs from west to east where $AB = 5$ miles. A boat leaves

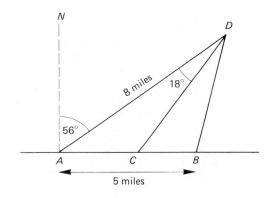

Fig. 12.44

A and sails 8 miles to *D* on a course 056° (N 56° E). The point *C* is between *A* and *B* such that ∠*ADC* = 18°. Calculate:
  (i) the distance *BD*;
  (ii) the distance *AC*.   [WJEC]

22 *A*, *B* and *C* are three points on level ground. *B* is due east of *A*, *C* is due north of *A*, and the distance *AC* is 160 m. The angles of elevation from *A* and *B* to the top of a vertical mast whose base is at *C* are respectively 30° and 28°. Calculate:
  (a) the distance *BC*;
  (b) the bearing of *C* from *B*.

23 The points *O*, *P*, *Q* and *T* lie in a horizontal plane, but, because of an obstruction, the point *T* is not visible from the point *O*. The bearings of the points *P* and *Q* from *O* are 029° and 062° respectively; *OP* = 0.8 km and *OQ* = 1.2 km. The bearing of *T* from *P* is 126° and the bearing of *T* from *Q* is 160°.
  Using a scale of 10 cm to 1 km, draw a diagram showing the points *O*, *P*, *Q* and *T*. From your diagram, obtain the bearing and distance of *T* from *O*.
                                                                              [O & C]

24 Figure 12.45 shows a storage shed on a building site.
  (a) Calculate the angle of inclination of the roof to the horizontal.
  (b) Calculate the length of the longest rod that can be stored in the shed.

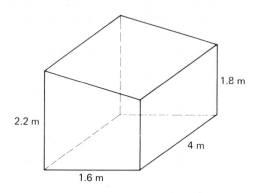

1.8 m
2.2 m
4 m
1.6 m

Fig. 12.45

25 The diagram in Fig. 12.6 shows a pyramid whose horizontal base is a square *ABCD*. The vertex *V* is vertically above *N*, the centre of the base.

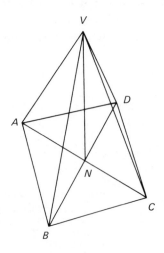

Fig. 12.46

(a) Given that the base has an area of 50 cm², find the length of a diagonal of the square *ABCD*.

(b) Given also that each slant edge of the pyramid is 13 cm, find the volume of the pyramid. [EAUC]

26 A square pyramid has base *A'B'C'D'* of side 50 cm. The top is cut off to give a square top *ABCD* of side 20 cm, height 1.5 m. Each of the four sloping sides is a trapezium of the same dimensions as *ABB'A'*. Find:

(a) the inclination of face *ABB'A'* to the base;

(b) the length of *AA'*;

(c) inclination of *AA'* to the horizontal.

27 In Fig. 12.47 *G* and *D* represent the positions of guns, *T* the position of a target and *O* that of an observation post.

Given that *OG* = 3500 m, *OT* = 2000 m, *OD* = 4000 m, angle *AOG* = 63°, angle *TAG* = 90° and *DOG* and *AOT* are straight lines, calculate the distances (a) *AG*, (b) *AT*, (c) *TD*. [EAUC]

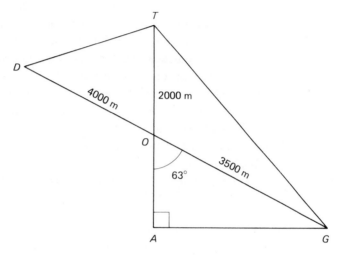

Fig. 12.47

28 The sails of a yacht form a triangle *ABC* with *BC* horizontal and *A* the top of the mast. The bottom of the mast *D* is the foot of the perpendicular from *A* to *BC*. *AD* = 6.5 m, *DC* = 2 m, *BD* = 3 m. When the mast is inclined at an angle 10° to the vertical, calculate the height of *A* above the water level (assumed to be the level of *BC*). When the mast is vertical the sun shines from the direction perpendicular to *BC* and at an elevation of 60°, calculate the area of the shadow cast by the sails on the water.

29 In triangle *ABC*, *BC* = 6.9 cm, ∠*BAC* = 28.9° and ∠*ABC* = 90°. Calculate the radius of the circumscribing circle of the triangle *ABC*.

30 In Fig. 12.48 *ABC* is a straight line, *CD* = 6.4 cm and *AB* = 7.3 cm. Calculate

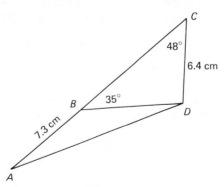

Fig. 12.48

(a) BD; (b) AD; (c) the area of triangle ABD; (d) the perpendicular distance of B from AD.

31 In a triangle ABC, $AB = 5$ cm, $\angle CAB = 68°$ and $\angle CBA = 30°$. D is the point on the opposite side of BC to A such that $\angle CBD = 65°$ and $BC = CD$. Calculate the lengths of BC and BD, correct to two decimal places.

32 In triangle PQR, $PQ = 5$ cm, $QR = 4$ cm, and $\angle QPR = 40°$. Calculate the two possible values of $\angle QRP$ and the corresponding length of PR. Illustrate your answer with a diagram. If only one triangle was possible, calculate the value of QR.

33 From a lighthouse L a boat is seen at a point A which is due north of L, and 1.2 km away from it. The boat travels in a straight line, and 10 minutes later it is at a point B on a bearing of $105°$ from L and 1.8 km away from L. Calculate:
(a) the speed of the boat in km/h;
(b) the direction in which the boat is travelling;
(c) the shortest distance between the boat and the lighthouse.

34 On a certain golf course, the distance from the tee T to the hole H is 350 m. A player drives his ball from T to a point S where $TS = 160$ m and $\angle STH = 3°$.
   Calculate, correct to 0.1 m, the distance SH.
   A second player drives his ball from T to a point O, such that $\angle HTO = 6°$ and $\angle TOH = 160°$. Calculate, correct to 1 m, the distance TO. A third player drives his ball a distance of 180 m from T to a point P which is 180 m from H. Calculate the angle PTH.

35 ABC is a triangle in which $AB = 4$ cm, $BC = 6.7$ cm and $\angle ABC = 105°$. Calculate AC.
   AB is produced to D so that $\angle BDC = 60°$. Calculate BD. Calculate also the area of the triangle ACD.

36 In triangle XYZ, $XY = 8.2$ cm and $YZ = 4.9$ cm. If $4.2 \leqslant XZ \leqslant 10.9$ and $P \leqslant \angle XYZ \leqslant q$, find p and q.

37 The points A and B lie on the circumference of a circle, of centre O and radius 49 cm, such that $\angle AOB = 80°$. Taking $\pi = \frac{22}{7}$, find the difference in length between the minor arc AB and the chord AB.

38 O is the centre of a circle of diameter 12 cm. P and Q lie on the circle and $\angle POQ = 80°$. Calculate:
(a) the length of the minor arc PQ;
(b) the area of the minor sector POQ;
(c) the area of the triangle POQ;
(d) the length of the chord PQ.

39 Fig. 12.49 shows the vertical cross-section of an oil drum, being fastened to the ground by a wire $PQRS$. The radius of the drum is 0.6 m. Calculate:
(a) RS;
(b) PS;
(c) angle ROQ (where O is the centre of the circle);
(d) the total length of the wire $PQRS$.

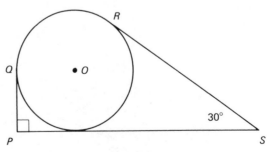

Fig. 12.49

40 State the latitude and longitude of the point A on the surface of the earth, if AB passes through the centre of the earth and B is ($56°$ N, $37°$ W).

41 On a globe whose diameter is 21 cm, two points on the same parallel of latitude differ in longitude by $120°$. The distance between the two points measured along the line of latitude is 5.5 cm. Calculate the latitude of the two points ($\pi = \frac{22}{7}$).

42 P and Q are two points on the same parallel of latitude $64°\ 25'$ S, whose longitudes differ by $180°$. Assuming the earth to be a sphere with centre O and radius 6370 km, calculate in kilometres:
(a) the radius of the parallel of latitude;
(b) the distance of P from Q measured along the parallel of latitude.
Calculate the shortest distance in nautical miles between P and Q measured on the surface of the earth.

43 A model globe has a radius of 10 cm and is marked with circles of latitude and longitude. It rests in a circular hole on a horizontal board in such a way that the circle of latitude $60°$ S is in contact with the board. Calculate:
(a) the diameter of the hole;
(b) the vertical height of the equator above the board;
(c) the fraction of each 'circle of longitude' which is above the board;
(d) the angle of the 'parallel of latitude' which is 6 cm vertically below the 'North Pole'.

44 A ($28°$ N, $162°$ W) and B ($28°$ N, $18°$ E) are two places on the earth's surface. Calculate in nautical miles, the distance between A and B:
(a) travelling by the shortest route;
(b) travelling along a parallel of latitude.
Two jet aircraft leave A simultaneously, one along each of the two different routes. Calculate the latitude and longitude of the other when the first aircraft arrives at B.

45 Figure 12.50(a) represents a strip of sticky tape of uniform width laid across a sheet of card near to one corner of the sheet which is right-angled. The strip is

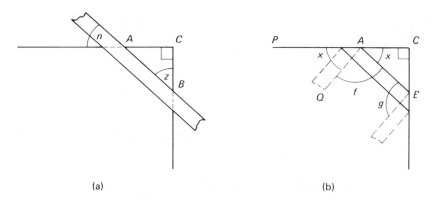

(a)   (b)

Fig. 12.50

then neatly folded under the sheet and pressed to the sheet, such that angle PAQ = angle BAC, as shown in (b). Similarly, the tape is folded at B so that equal angles are formed with the edge of the card.
(a) If $x = 37°$, calculate the sizes of the angles:
    (i) n; (ii) z; (iii) f; (iv) g.
(b) (i) Find a formula for f in terms of x;
    (ii) find a formula for g in terms of x;
    (iii) for what values of x are the folded parts of the tape parallel?

[UCLES]

13 Drawing and Construction

13.1   Constructions Using Ruler and Compass

In the following diagrams, points numbered ①, ② etc. show where and the order in which the compass point is placed and the corresponding arcs have the same numbers. Notes about restrictions on the radii are given at the side.

(a) Bisecting a Line *PQ* at Right Angles

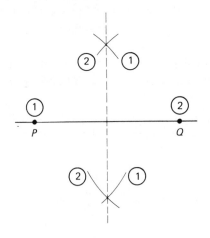

Fig. 13.1

Radius ① = radius ②, and must be greater than half of *PQ*.

(b) Bisecting an Angle *ABC*

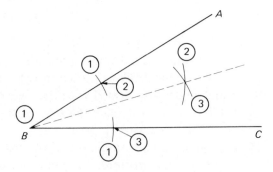

Fig. 13.2

① Any radius;
radius ② = radius ③, large enough for two arcs to meet.

(c) Perpendicular from a Point P to a Line AB

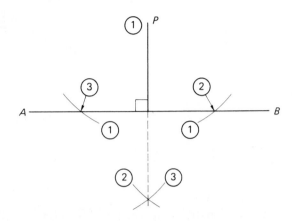

Fig. 13.3

Radii ①, ②, ③ equal, large enough for ① to meet line AB.

(d) Construction of 60° at a Point X on XY

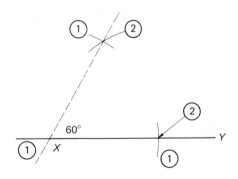

Fig. 13.4

Radius ① = radius ②, any length.

(e) Construction of 30°, 45°, 75° etc.

30° can be obtained by bisecting 60°.
45° can be obtained by bisecting 90°.
75° can be obtained by constructing 45°, and then 30° on this.

Many other angles could be constructed using these basic angles,

e.g. $52\frac{1}{2}° = 15° + 37\frac{1}{2}° =$ half of 30° + half of 75°,
or　　　$=$ half of 105° $=$ half of (90° + 15°).

(f) Drawing a Line Through *P* Parallel to *AB*

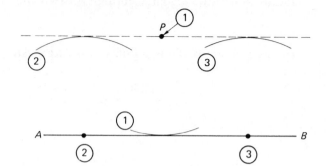

Fig. 13.5

Radii ①, ②, ③ equal, the arc ① just touches *AB*.

(g) To Construct an Angle of 90° at *A* on *AB*

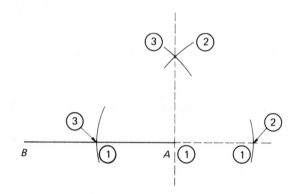

Fig. 13.6

BA must be produced at *A*;
radii ①, ②, ③ equal.

(h) To Construct the Circumscribing Circle of a Triangle *ABC*

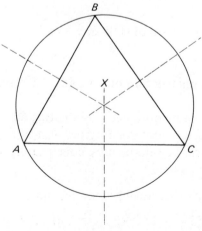

Fig. 13.7

To find the centre of the circle, bisect each side in turn. The intersection of the bisectors is the centre of the circle.

(i) To Construct the Inscribing Circle of Triangle *PQR*

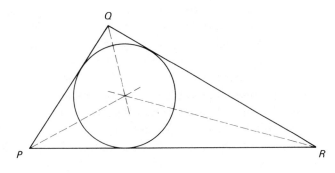

Fig. 13.8

The three angles are bisected. The point of intersection of the bisectors of the angles is the centre of the circle.

(j) Division of a Line in a Given Ratio

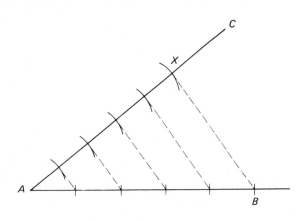

Fig. 13.9

Strictly speaking, this is not a ruler and compass construction.

Divide a line *AB* into 5 equal parts.

Draw line *AC* (any reasonable angle). Use a compass to mark off 5 equal lengths on *AC*. Join the end point *X* to *B*.

Then using two set squares, draw 4 lines parallel to *BX*. *AB* is then divided into 5 equal parts.

In all construction work, the construction lines must be faint, and not detract from the main details of the diagram.

13.2 Velocity Triangles

The work on velocity triangles covered in Chapter 7 and Chapter 12 can be used when solving the problems by scale drawing.

Always state clearly what scale you are using when drawing the diagrams.

Worked Example 13.1

Three towns O, A and B are such that O is the point $(0, 0)$, $\vec{OA} = \begin{pmatrix} 8 \\ -2 \end{pmatrix}$ and $\vec{OB} = \begin{pmatrix} 24 \\ 6 \end{pmatrix}$, where the units are kilometres.

(a) Using a scale of 1 cm to 2 km, illustrate O, A and B on squared paper.

(b) A motorway is constructed so that it is equidistant from A and B. An interchange on this motorway is to be located so that it is not more than 10 km from A. Assuming that the motorway can be represented by a straight line, accurately construct and label on your diagram the line segment p which represents possible locations for the interchange.

(c) Circle and label the point X on p which is closest to A. Calculate, in kilometres, the distance AX, giving your answer correct to two significant figures.

(d) A new town C is to be situated so that it is equidistant from the motorway and the line AB. Mark clearly on your diagram all the possible locations for C if $CX = 8$ km.

[AEB]

Solution

(a), (b) The diagram in Fig. 13.10 shows accurately the position P_1 and P_2 at the end of the line segment p. These are found by drawing a circle, centre A, radius

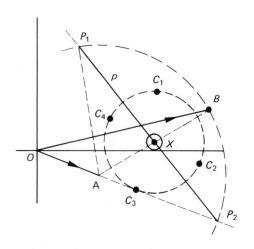

Fig. 13.10

10 km (5 cm), and finding the intersection of it with the perpendicular bisector of AB. (So that the motorway is equidistant from A and B.)

(c) The closest point X is the mid-point of AB, hence X is $(16, 2)$.
   AX can be found by Pythagoras' theorem.
   The horizontal distance from A to X is 10.
   The vertical distance from A to X is 4.
   $\therefore AX = \sqrt{10^2 + 4^2} = 10.8$ km

(d) Draw a circle centre X radius 8 km (4 cm). The new town must lie on the bisectors of angle $P_1 XB$ and $P_2 XB$. (These have been left out for clarity.) We thus obtain four points, C_1, C_2, C_3, C_4 as shown.

Worked Example 13.2

Using ruler and compasses throughout, construct a rectangle *PQRS* in which *PQ* = 10 cm and the diagonal *QS* = 11 cm. Measure and state the length of PS.

Construct a triangle *TPQ* having *PQ* as base, the point *T* on *SR* and the angle *TQP* = 45°. Measure and state the length of *PT*. Find, by construction, a point *X* on *SR* between *S* and *T* and such that the angle *PXQ* = 90°. Complete the parallelogram *PXYQ*.

Without any further measurement, calculate:
(a) the area of the triangle *TQP*;
(b) the area of the parallelogram *PXYQ*;
(c) the product of *PX* and *QX*. [AEB]

Solution

Draw *PQ* and mark off 10 cm. Construct 90° at *P*. Mark off *QS* = 11 cm. Repeat to find *R*.

By measurement, *PS* = 4.6 cm.
Bisect ∠*RQP* to get 45°; locate *T*, join *TP*.
By measurement, *PT* = 7.9 cm.

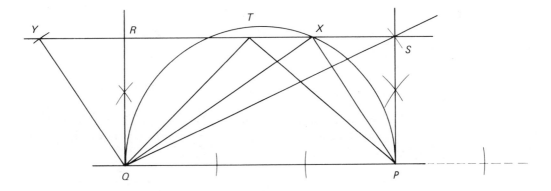

Fig. 13.11

To find *X*, *QP* will be the diameter of a circle if *QXP* = 90°. Hence, draw a semicircle radius 5 cm, centre mid-point of *QP*, to cut *RS* at *X*.

Y can be found by marking off 10 cm.
(a) Area of △*TQP* = ½ × 10 × 4.6 = 23 cm².
(b) Area of parallelogram = 10 × 4.6 = 46 cm².
(c) Since *PXQ* is a right-angled triangle, *PX* multiplied by *QX* is twice the area of the triangle. But △*PXQ* and △*PTQ* have the same area.
∴ *PX* · *QX* = 46.

Worked Example 13.3

A circle is inscribed inside a square *ABCD*. Any point *P* of the circle is joined to any point *Q* of the square and *PQ* is produced to *R*, so that *PR* = 2*PQ*. Describe and illustrate:
(a) the locus of *R* as *Q* moves round the square for any fixed point *P* not a point of contact;

(b) the locus of R as P moves round the circle for any fixed point Q not a point of contact;

(c) the locus of R for one of the points of contact of the circle with the square.

[L]

Solution

This is quite a difficult question, and an explanation is not easy to give. The approach will be to look at several positions of R for each case and see how this leads to a solution.

(a) Figure 13.12 shows five different positions of Q and the corresponding positions of R. Since Q is the mid-point on each occasion, the locus is a square with sides parallel to the original square and sides of twice the length. The centre of enlargement will be P.

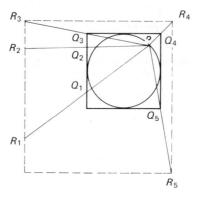

Fig. 13.12

(b) This time three positions of P are given. The resulting locus is an enlargement, scale factor -1, centre of enlargement Q. It is another circle.

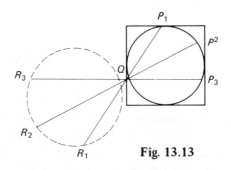

Fig. 13.13

(c) Two positions of Q are shown; it can easily be seen that the locus is another circle which is an enlargement, scale factor 2, centre of enlargement P.

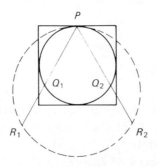

Fig. 13.14

13.3   Plans and Elevations

The solid shown in Fig. 13.15 is similar to the shape of a house. Hidden edges are dotted. Three different views of it are shown alongside.

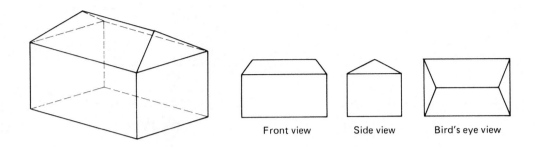

Front view   Side view   Bird's eye view

Fig. 13.15

The front and side views are referred to as *elevations*, and the bird's eye view is called the *plan*. These can be combined together in a more technical fashion as shown in Fig. 13.16.

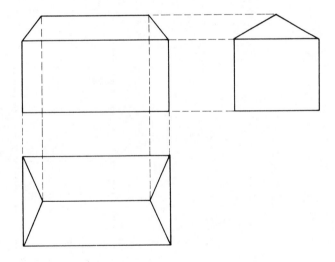

Fig. 13.16

Diagrams of this nature are commonly used in engineering. They are very useful for finding distances through three-dimensional objects, as illustrated by the following example.

Worked Example 13.4

Figure 13.17 shows a framework made from 12 rods of length 8 cm, forming a cube. $A'X = 3$ cm. Draw a plan and elevation from the direction of the arrow, and hence find the length of DX.

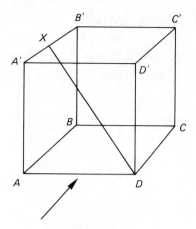

Fig. 13.17

Solution

The plan and elevation are shown in Fig. 13.18. In order to find the real distance *DX*, read the *horizontal* distance from the plan, and the *vertical* distance from the elevation.

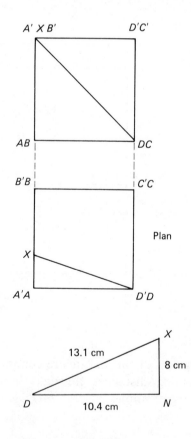

Fig. 13.18

The triangle *XDN* can then be drawn, where *N* is the point vertically beneath *X*.

The actual distance *XD* = 13.1 cm.

13.4   Locus

As with velocity triangles, the work covered so far on locus in Chapters 6 and 7 can be used here.

Worked Example 13.5

AB is a fixed line of length 6 cm. Construct the locus of a variable point *P* such that $\angle APB = 45°$.

Solution

If $\angle APB$ is always $45°$, then *P* moves along the arc of a circle as shown in Fig. 13.19. To find the centre of the circle, one position of *P* must be found. The easiest case is if $\hat{A} = 45°$ and $\hat{B} = 90°$.

(a) Construct $\hat{B} = 90°$.
(b) Bisect $90°$ at *A*.
(c) Bisect the sides to find the centre of the circle.
(d) Draw the arc. See Fig. 13.20.

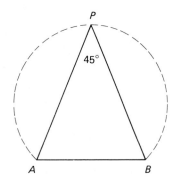

Fig. 13.19

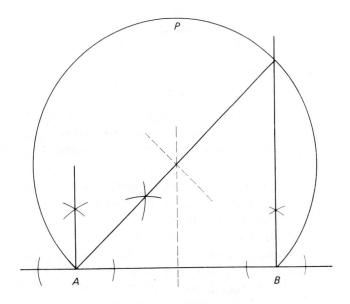

Fig. 13.20

Exercise 13

This exercise does not contain multiple-choice questions as they are not relevant.

1 State the locus in two dimensions of the centre of a variable circle:
   (a) which touches a given line at a given point;
   (b) which touches both of two fixed parallel lines.
   On graph paper, draw the rectangle $ABYX$ in which $AB = 10$ cm, $BY = 8$ cm. Draw also the line LM where L is the mid-point of AX and M is the mid-point of BY.
   Determine, by construction, the point P lying inside the rectangle $ABYX$, such that it is equidistant from AB and LM, and $AP = 5$ cm.
   Construct a circle which passes through P touching AB and XY.
   Construct a further circle which touches the circle you have drawn and also the lines LM and XY. Label its centre R and measure and write down the length of PR. [L]

2 In Fig. 13.21 construct, using ruler and compasses only, and leaving all your construction lines clearly visible:

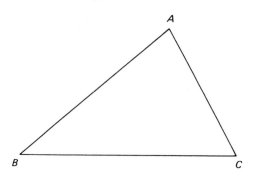

Fig. 13.21

   (a) the perpendicular bisector of BC;
   (b) the bisector of angle ABC. D is the point at which the two bisectors meet.
   (c) Measure the length of DC.
   (d) Shade, with horizontal lines, the set of points P, inside triangle ABC, for which $PB \leqslant PC$.
   (e) Shade, with vertical lines, the set of points Q, inside triangle ABC, for which angle $QBC \leqslant \frac{1}{2}$ angle ABC. [OLE]

3 Using ruler and compasses only, construct in a single diagram:
   (a) the triangle ABC such that $AB = 7$ cm, $AC = 10$ cm and $BC = 6$ cm;
   (b) the circle centre P which touches AC and also touches CB produced and is such that $CP = 8$ cm;
   (c) the triangle QAB, equal in area to the triangle ABC, where QA is perpendicular to AC and Q is on the opposite side of AC to B.
   State the length of CQ. [AEB]

4 Draw a circle of radius 5 cm. Using ruler and compasses only construct a cyclic quadrilateral $ABCD$ having its vertices on this circle and having $DC = 8$ cm, the angle $ADC = 60°$ and $AB = BC$ and B and D lying on opposite sides of AC. Construct a triangle ADX, with X on DC produced, equal in area to the quadrilateral $ABCD$. Mark the point X clearly on the diagram and measure AX. [AEB]

5 Using ruler and compasses only, construct the triangle ABC in which AB = 8 cm, BC = 9.2 cm, and $\angle ABC = 120°$. Construct also (a) the perpendicular from B to AC; (b) the bisector of angle BAC.

6 Three equal spheres stand on a plane so that they touch each other. An identical sphere is placed on top. Draw the plan of this configuration.

7 Draw a line XY of length 7 cm. Construct a line PQ parallel to XY, and distant 4 cm from XY. Construct the circle which passes through X and Y, touching PQ. Measure the radius.

8 Draw any acute-angled triangle BAC. Produce BC to D where $CD = CA$. Produce CB to E where $BE = BA$. Construct the bisectors of angles ABE and ACD. Label the point of intersection of these two bisectors O. By drawing a circle through A with centre O, deduce that OB is the line of symmetry of the quadrilateral $OABE$. Construct also the bisector of angle BAC. What do you notice?

9 Construct a hexagon of side 3 cm.

10 Describe the locus, in a plane, of
   (a) the vertex A of $\triangle ABC$ when BC is fixed in length and position and $\angle BAC$ is of constant size;
   (b) the vertex P of $\triangle PRQ$ when QR is fixed in length and position and $\triangle PQR$ is of constant area.
      Using ruler and compasses only and drawing any necessary area of such length and clarity to permit assessment, construct a $\triangle XYZ$ in which YZ = 8 cm, $\angle YXZ = 60°$ and the area of $\triangle XYZ = 24$ cm^2. In the case when XY is longer than YZ, measure and write down
   (c) the length of XY to the nearest mm;
   (d) the $\angle XYZ$ to the nearest degree. [L]

11 The $\triangle PQR$ lies in a fixed plane, has a fixed base QR and a constant area. State the locus of the vertex P.
   Using ruler and compasses only in the remainder of this question, and drawing any necessary arcs of such length and clarity to permit assessment, construct:
   (a) a $\triangle ABC$ in which AB = 8 cm, BC = 6 cm and AC = 7 cm;
   (b) the circumcircle of $\triangle ABC$;
   (c) the cyclic quadrilateral $BCAX$, where X is such that area $\triangle BCX$ = area $\triangle BCA$;
   (d) $\triangle XBY$, with Y on BC produced, which is such that the area of $\triangle XBY$ is equal to the area of quadrilateral $BCAX$. Measure, and write down, the length of XY. [L]

12 Figure 13.22 illustrates a model of a building, made from two solid blocks of

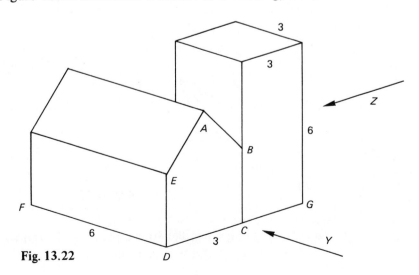

Fig. 13.22

231

wood fixed together, standing on a horizontal table. One block is rectangular; its height is 6 cm and its cross-section is a square of side 3 cm. The other block is a prism of length 6 cm with cross-section $ABCDE$. The vertical height of A above DC is 5 cm, $ED = BC = DC = 3$ cm, $EA = AB$ and $\angle EDC = \angle DCB = 90°$.

Draw full size:

(a) the plan of the model;

(b) the elevation, as viewed from Y, on a vertical plane parallel to the straight line DCG;

(c) the elevation, as viewed from Z, on a vertical plane parallel to DF.

Find the volume of wood used to make the model. [UCLES]

13 Two points P and Q are 6 cm apart.

(a) Construct the locus of points R such that $AR = 2.5$ cm.

(b) Construct the locus of points R such that $\angle ARB = 90°$.

(c) Construct the locus of points R such that $\angle ARB = 60°$.

14 A plot of land on a proposed housing estate is in the shape of a triangle ABC with $AB = 12.5$ m, $AC = 9$ m and the angle $BAC = 60°$. There is a preservation order on a tree T on the plot and T is equidistant from CA and CB and is also 8 m from B.

Using ruler and compasses only, draw a scaled diagram of this plot of land and by construction show the position of T.

Measure and state the distance AT.

An architect's plan of this housing estate is drawn to a scale of 1 : 250. Calculate the length on the architect's plan, in cm, of the side AB.

The area of a garden is represented by 44 cm² on the plan. Calculate, in m², the actual area of the garden. [AEB]

15 Construct the triangle XYZ in which $XY = 5$ cm, $\angle X = 60°$ and $\angle Y = 90°$. Measure and write down the length of YZ. On the same diagram,

(a) Construct the circumcircle of $\triangle XYZ$.

(b) Construct, on the same side of XY as Z, the locus of the point P such that the area of $\triangle XYP$ equals half the area of $\triangle XYZ$.

(c) Mark, and label clearly, a point Q such that $\angle XQY = 30°$ and the area of $\triangle XYQ$ is half the area of $\triangle XYZ$.

Given that M is a point such that $\angle XMY = 30°$, find the largest possible area of $\triangle XMY$. [UCLES]

16 Figure 13.23, diagram I, shows a bowl in which rests a stiff rectangular card $ABCD$. The bowl is a thin hollow hemisphere of radius 62 mm with part of it

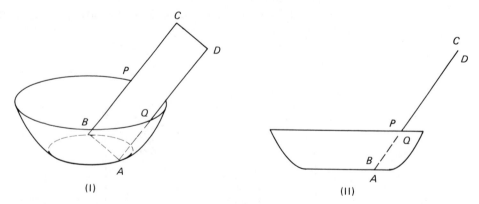

Fig. 13.23

cut off by a plane parallel to the plane of the rim to form the base. The bowl is 49 mm deep. Diagram II is an elevation, in which the card is seen edge-on.

(a) Draw, full-size, the elevation of the bowl without the card. Measure the diameter of the base and draw, full-size, the plan of the bowl.

(b) The area of the outer curved surface can be found by multiplying the depth by the circumference of the rim. Calculate the area of the whole of the outer surface, including the base.

(c) The corners A and B of the card are on the circumference of the base and the edges BC and AD rest on the rim at P and Q. The length AB is 58 mm. Show $ABPQ$ first in your plan and then in your elevation. Measure the inclination of the card to the base.   [O & C]

17 Use a ruler and protractor to answer this question, but do not use graph paper.

The front view of a large house is a rectangle with a trapezium-shaped roof on top.

The width of the house is 20 metres and its height, to the start of the roof, is 16.8 metres.

The sloping sides of the symmetrical roof are 9 metres long, at an angle of 35° to the horizontal (inwards).

Using the scale of 1 cm = 2 m, draw a diagram which represents the front view of the house.

Use your drawing to find:

(a) the actual length of the horizontal part of the top of the roof;

(b) the overall height of the house;

(c) the volume of roof space.

14 Abstract Algebra

14.1 Binary Operations

Most people use binary operations without realizing it. Addition is a binary operation, because if it is applied to two numbers, it produces a single answer. Any mathematical operation which acts on two quantities to produce a single answer is called a binary operation. In this chapter, we will investigate the properties of a number of such operations.

We now give some examples.

(a) If $a * b$ is the average of a and b,

then, $2 * 1 = \dfrac{2+1}{2} = 1.5$;

$1 * 2 = \dfrac{1+2}{2} = 1.5$;

$(4 * 5) * 6 = \left(\dfrac{4+5}{2}\right) * 6 = 4.5 * 6 = \dfrac{4.5+6}{2} = 5.25$;

$4 * (5 * 6) = 4 * \left(\dfrac{5+6}{2}\right) = 4 * 5.5 = 4.75$.

It follows that $a * (b * c) \neq (a * b) * c$, so that $*$ is not *associative*. However, $a * b = b * a$, so that $*$ is *commutative*.

(b) If $x \dagger y$ is the LCM of x and y

then $8 \dagger 12 = 24$ and $12 \dagger 8 = 24$;

$(8 \dagger 16) \dagger 12 = 16 \dagger 12 = 48$;

$8 \dagger (16 \dagger 12) = 8 \dagger 48 = 48$.

Hence it appears that $*$ is associative (although this does not constitute a proof) and commutative.

(c) If $p \circ q = p + q - pq$,

then $(2 \circ 3) = 2 + 3 - 6 = -1$;

$(3 \circ 2) = 3 + 2 - 6 = -1$;

$(2 \circ 3) \circ 4 = -1 + 4 + 4 = 7$;

$2 \circ (3 \circ 4) = 2 \circ (3 + 4 - 12) = 2 \circ -5 = 2 - 5 + 10 = 7$.

If would appear that $\circ$ is associative and commutative.

14.2 Closure

If a binary operation $*$ is used on a particular set S, and whatever two elements of S are chosen, the answer is always a member of S, then S is said to be closed under $*$. In Fig. 14.1, the table is closed.

14.3 Identity

A set S is said to have an identity element $e \in S$ under the operation $*$, if for all $x \in S$, $x * e = e * x = x$.

If S is the set of integers, the identity under $+$ is 0, since $x + 0 = 0 + x = x$ always.

14.4 Inverse

The inverse of an element x, denoted by x^{-1} (this does not necessary mean $1/x$) has the property that under an operation $\circ$,

$$x \circ x^{-1} = x^{-1} \circ x = e.$$

The inverse of 4 under the operation of addition is -4, since the identity element is 0, and $4 + -4 = 0$.

14.5 Operation Tables

An operation on a set can also be defined by means of a table, called an *operation table*, which gives all possible effects of the operation on any two elements of the set. Consider the table shown in Fig. 14.1.

$*$	e	a	b	c
e	e	a	b	c
a	a	e	c	b
b	b	c	e	a
c	c	b	a	e

Fig. 14.1

The entry marked with an arrow in row 4, column 3, stands for

$$c * b.$$

The order of letters is crucial:

c is the element at the beginning of row 4;

b is the element at the beginning of column 3.

We have, then, $c * b = a$.

The identity is e, because the row with e at the front leaves all letters unchanged. To find the inverse say of b, look along the row beginning b until you come to e, look at the top of the column, and you find b.

Hence, $b^{-1} = b$. This is an example of a *self inverse* element.

In fact, $a^{-1} = a$ and $c^{-1} = c$.

The *leading diagonal* of the table contains only the identity e, and it can be seen that the table is symmetrical about this diagonal. This means that the operation is commutative. You should now try the following straightforward exercise on abstract operations.

Check 30

1 If $p * q = 2p + 3q$, find $3 * 4$ and $4 * 3$. What can you say about $*$?
2 If $a * b = a^2 - b^2$, find $4 * 5$.
3 If $a * b = a^2 + b^2 + 2ab$, gives an example to show that $*$ is not associative.
4 If $a * b = a/b$, which of the following are true?

(a) $a * b = b * a$ (b) $a * (b * c) = \dfrac{ac}{b}$

(c) $(a + b) * c = a * c + b * c$ (d) $c * (a + b) = c * a + c * b$.

5 If $a * b = \dfrac{a}{a + b}$, find (a) $1 * 2$ (b) $2 * 1$ (c) $(1 * 1) * 1$

6 If $agb = \sqrt{a^2 + b^2}$, find
    (a) $3g4$ (b) $(3g4)g12$ (c) $3g(4g12)$

14.6 Groups

The idea of a set, together with an operation, is known as an algebraic structure. There are many in mathematics, but this section deals with one in particular called a group. We given here a formal definition of a group.

> The set $G = \{a, b, c, \ldots\}$ with an operation $\circ$ is a *group* if
> (a) For all $a, b \in G$ $a \circ b \in G$ (i.e. closed).
> (b) The operation $\circ$ is associative.
> (c) One element of G is an identity.
> (d) All elements of G have an inverse that is in G.

These are not very difficult at first sight, although proving whether $\circ$ is associative is often too difficult at this level.

Worked Example 14.1

A binary operation $*$ is defined on the set

$$S = \{(r, \theta) : r \geqslant 0, 0 \leqslant \theta < 360\}$$

such that $(a, b) * (c, d) = (ac, b \oplus d)$ where $\oplus$ represents addition modulo 360.
(a) Evaluate $(2, 120) * (3, 300)$.
(b) Evaluate $(a, b) * (1, 0)$. What does this suggest about $(1, 0)$?
(c) Evaluate $(p, q) * \left(\dfrac{1}{p}, 360 - q \right)$ and hence write down the inverse of $(\frac{2}{3}, 170)$.
(d) Find r and θ if $(r, \theta) * (3, 70) = (6, 30)$.
(e) Find r and all possible values of θ if

$$[(r, \theta) * (r, \theta)] * (r, \theta) = (8, 0).$$ [AEB]

Solution

(a) $(2, 120) * (3, 300) = (2 \times 3, 120 \oplus 300)$
                      $= (6, 60)$, since $420 \equiv 60 \pmod{360}$.
(b) $(a, b) * (1, 0) = (a \times 1, b \oplus 0)$
                   $= (a, b)$.
    This suggests that $(1, 0)$ is the identity element under $*$.

(c) $(p, q) * \left(\dfrac{1}{p}, 360 - q\right) = \left(p \times \dfrac{1}{p}, q \oplus 360 - q\right)$

$\qquad\qquad\qquad\qquad = (1, 360)$ but $360 \equiv 0 \pmod{360}$

$\qquad\qquad\qquad\qquad = (1, 0).$

This means that $\left(\dfrac{1}{p}, 360 - q\right)$ is the inverse of (p, q),

$\therefore$ the inverse of $(\frac{2}{3}, 170)$ is $(\frac{3}{2}, 190)$.

(d) If $(r, \theta) * (3, 70) = (6, 30),$

then $(3r, 70 + \theta) = (6, 30).$

Clearly $r = 2,$

but if $70 + \theta = 30,\ \theta = -40.$ This is not in the range of values of θ, hence 360 must be added.

$\therefore \theta = 320.$

(e) $(r, \theta) * (r, \theta) = (r^2, 2\theta);$

$(r^2, 2\theta) * (r, \theta) = (r^3, 3\theta);$

$\therefore r^3 = 8 \quad \Rightarrow \quad r = 2$

and $3\theta \equiv 0 \pmod{360}, \therefore \theta = 0, 120, 240$ [since $720 \equiv 0 \pmod{360}$].

Worked Example 14.2

A binary operation $*$ is defined on all the non-zero rational numbers by $x * y = \dfrac{xy}{x + y}$, where x and y are any non-zero rationals and $x + y \neq 0.$

(a) Evaluate $2 * 3$ and $\frac{2}{3} * \frac{3}{2}$, leaving your answers as fractions.

(b) Evaluate $(2 * 3) * 4$ and $2 * (3 * 4)$. What do your results suggest about the operation $*$?

(c) Solve the equation $x * x = 5.$

(d) Solve the equation $(x * x) + (1 * x) - 1 = 0.$

(e) Prove that there is no non-zero rational number which can act as an identity element for the operation. [AEB]

Solution

(a) $2 * 3 = \dfrac{2 \times 3}{2 + 3} = \dfrac{6}{5}$

$\dfrac{2}{3} * \dfrac{3}{2} = \dfrac{\frac{2}{3} \times \frac{3}{2}}{\frac{2}{3} + \frac{3}{2}} = \dfrac{1}{\frac{13}{6}} = \dfrac{6}{13}.$

(b) $(2 * 3) * 4 = \dfrac{6}{5} * 4 = \dfrac{\frac{6}{5} \times 4}{\frac{6}{5} + 4} = \dfrac{\frac{24}{5}}{\frac{26}{5}} = \dfrac{12}{13}.$

$2 * (3 * 4) = 2 * \left[\dfrac{3 \times 4}{3 + 4}\right] = 2 * \dfrac{12}{7} = \dfrac{2 \times \frac{12}{7}}{2 + \frac{12}{7}} = \dfrac{\frac{24}{7}}{\frac{26}{7}} = \dfrac{12}{13}.$

This suggests that $*$ is associative.

(c) Always change abstract equations into a proper equation by using the definition.

Hence if $x * x = 5,$

this means $\dfrac{x \times x}{x + x} = 5,$ $\qquad\qquad \therefore \dfrac{x^2}{2x} = 5.$ $\qquad\qquad\qquad$ (1)

The original definition did not allow $x + y = 0$,

∴ it doesn't allow $x + x = 2x = 0$, i.e. $x = 0$.

We can divide (1) by x then, to get

$$\frac{x}{2} = 5, \therefore x = 10.$$

(d) Similarly, here, we have

$$\frac{x \times x}{x + x} + \frac{1 \times x}{1 + x} - 1 = 0,$$

$$\therefore \frac{x^2}{2x} + \frac{x}{x + 1} - 1 = 0.$$

Multiply by $2(x + 1)$:

$$\cancel{2}(x + 1) \times \frac{x}{\cancel{2}} + 2\cancel{(x + 1)} \times \frac{x}{\cancel{(x + 1)}} - 2(x + 1) = 0$$

$$\therefore x^2 + x + 2x - 2x - 2 = 0,$$

$$\therefore x^2 + x - 2 = 0,$$

$$\therefore (x + 2)(x - 1) = 0,$$

$$\therefore x = 1 \text{ or } -2.$$

(e) If an identity e exists, then

$x * e = x$,

$$\therefore \frac{\cancel{x}e}{x + e} = \frac{1}{\cancel{x}}; \text{ since } x \neq 0, \text{ divide by } x:$$

$$\therefore e = x + e.$$

This gives no value of e. Hence an identity does not exist.

Worked Example 14.3

Functions e, f and g which map the set $S = \{1, 2, 3\}$ to itself are defined as follows:

e = {(1, 1), (2, 2), (3, 3)},

f = {(1, 2), (2, 3), (3, 1)},

g = {(1, 3), (2, 1), (3, 2)}.

(a) Define each of the composite functions f ∘ g and (f ∘ f) ∘ f as a set of ordered pairs (f ∘ g means g followed by f).

(b) Copy and complete the composition table in Fig. 14.2.

∘	e	f	g
e			
f			
g	g		

Fig. 14.2

(c) Explain fully why the set {e, f, g} forms a group under the operation ∘. (You may assume associativity.) Another function h which maps S to itself is such that h ∘ h = e and h ∘ f = g ∘ h.

(d) Find single functions, belonging to the set {e, f, g, h}, which represent *each* of the composite functions.

$$h \circ f \circ h, (h \circ f \circ h)^{-1}, \text{ and } f \circ h \circ f \circ h.$$

(You may assume associativity.)

(e) Write down two sets, each containing two elements, which form groups under the operation $\circ$.　　　　　　　　　　　　　　　　　　　　　　　　[AEB]

Solution

(a) Consider the number 1.

　$g(1) = 3$ and $f(3) = 1, \therefore fg(1) = 1$.

　Similarly,

　$g(2) = 1$　　and $f(1) = 2$,　　$\therefore fg(2) = 2$;

　$g(3) = 2$　　and $f(2) = 3$,　　$\therefore fg(3) = 3$.

　hence $f \circ g = \{(1, 1), (2, 2), (3, 3)\}$,

　$f(1) = 2$,　　$f(2) = 3$,　　$f(3) = 1$,　　$\therefore (f \circ f) \circ f(1) = 1$;

　also, $fff(2) = 2$　　and $f^3(3) = 3$　　using all 3 notations,

　$\therefore (f \circ f) \circ f = \{(1, 1), (2, 2), (3, 3)\}$.

(b) When completing the table it is clear that e has no effect and is therefore the identity.

　We have $f \circ g = e$.

　The table now looks like Fig. 14.3.

o	e	f	g
e	e	f	g
f	f		e
g	g		

Fig. 14.3

Do not make any assumptions from now on.

$g \circ g = \{(1, 2), (2, 3), (3, 1)\} = f$

$f \circ f = \{(1, 3), (2, 1), (3, 2)\} = g$

$g \circ f = \{(1, 1), (2, 2), (3, 3)\} = e$.

　The table can now be completed; look at Fig. 14.4.

o	e	f	g
e	e	f	g
f	f	g	e
g	g	e	f

Fig. 14.4

(c) It is a group because

　(A) identity = e;

　(B) it is closed;

　(C) $f^{-1} = g, g^{-1} = f$;

　(D) it is associative.

(d) It is not necessary to find out explicitly what h is.

$$h \circ f \circ h = (h \circ f) \circ h = (g \circ h) \circ h = g \circ (h \circ h)$$
$$\underbrace{\qquad \text{Given} \qquad} \qquad = g \circ e \leftarrow \text{Given}$$
$$= g$$

$$\therefore (h \circ f \circ h)^{-1} = g^{-1} = f.$$
$$f \circ h \circ f \circ h = f \circ (g \circ h) \circ h = (f \circ g) \circ (h \circ h)$$
$$= e \circ e = e.$$

(e) The set must contain e. Since $h \circ h = e$, one set must be $\{e, h\}$.

Also, $f \circ h$ is self inverse, $\therefore$ another set would be $\{e, f \circ h\}$.

Proving $\circ$ is associative is often difficult. At this level you will probably be told whether or not the operation is associative. The table shown as Fig. 14.1 is in fact a group, but without checking all possible combinations of three elements, it is difficult to see that it is associative.

14.7 Symmetry Groups

Consider the rectangle shown in Fig. 14.5(a) where the corners have been labelled to help find positions.

Let M_x = reflection in Ox,
M_y = reflection in Oy,
R_1 = rotation of $180°$ and
R_2 = rotation of $360°$

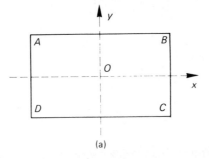

(a)

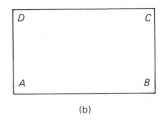

(b)

Fig. 14.5

The table is completed as in Fig. 14.6. The element ringed is $M_y * R_1$. This means R_1 followed by M_y. The new position of the rectangle is shown in Fig. 14.5(b).

*	M_x	M_y	R_1	R_2
M_x	R_2	R_1	M_y	M_x
M_y	R_1	R_2	(M_x)	M_y
R_1	M_y	M_x	R_2	R_1
R_2	M_x	M_y	R_1	R_2

Fig. 14.6

240

This could have been obtained in one go from M_x:

$$M_y * R_1 = M_x.$$

The table in Fig. 14.6 is called the *symmetry group* of the rectangle. Composition of transformations is associative.

14.8 Modulo Arithmetic

Consider the set $\{0, 1, 2, 3\}$ under addition + modulo 4. This means $3 + 2 = 5 \equiv 1$ (mod 4) (rather like clock arithmetic).

The set forms a group, with identity equal to 0 (Fig. 14.7).

+	0	1	2	3
0	0	1	2	3
1	1	2	3	0
2	2	3	0	1
3	3	0	1	2

Fig. 14.7

14.9 Subgroups

A subset of a group G, which is itself a group, is called a *subgroup* of G. See Worked Example 14.3.

14.10 Isomorphism

Consider the two groups in Figs 14.1 and 14.6. Apart from the letters used, the structures of the tables are identical. We say that the two groups are *isomorphic*. The best clue as to whether or not two groups are isomorphic is the leading diagonal. In Figs 14.1 and 14.6, the elements on the leading diagonal are all the same. In Fig. 14.7 they are not, hence Fig. 14.7 is not isomorphic to the other 2.

Check 31

Which of the following forms a group under the given operation?

1 $\{0, 1, 2, 3\}$ under the operation + modulo 4.
2 $\{1, 2, 3\}$ under the operation × modulo 4.
3 $\{1, 3, 5, 7, 9\}$ under the operation × modulo 10.
4 $\{1, 3, 5, 7\}$ under the operation × modulo 8.
5 $\{1, -1\}$ (a) under addition, (b) under multiplication.
6 $\{0, 1, 2, 3, 4, 5\}$ under the operation × modulo 6.
7 $\{1, 5\}$ under the operation × modulo 10.
8 $\{1, 5, 7, 11\}$ under the operation × modulo 12.
9 $\{\phi, \{a\}, \{b\}, \{a, b\}\}$ under the set operation $\cap$.
10 $\{\phi, \{a\}, \{b\}, \{a, b\}\}$ under the set operation $\cup$.

Exercise 14

1 If $a * b = 2a + 3b$, then $(1 * 2) * 3$ equals:
   A 41   **B** 25   **C** 11   **D** 8

2 If $x \dagger y = x + y + 2$, and $(x \dagger 1) \dagger 2 = 4$, then x equals:
   A -1 **B** 1   **C** -3 **D** 3

3 If $p * q$ is the remainder after pq is divided by 10, then $(p * q) * r$ equals:
   A $p * (q * r)$ **B** $\dfrac{pqr}{10}$ **C** $\dfrac{pqr}{100}$ **D** none of these

4 For which of the following definitions of $*$ on the set of non-zero real numbers is $x * y$ not equal to $y * x$?
   A $x * y = x^2 + y^2$ **B** $x * y = x + y$

   C $x * y = 2(x + y)$ **D** $x * y = \dfrac{x}{y} + \dfrac{y}{x}$ **E** $x * y = x - y$ [SCE]

5 If $a * b$ denotes the average of a and b, then $12 * (8 * 3) - (12 * 8) * 3$ equals:
   A 0   **B** $6\frac{1}{2}$ **C** $2\frac{1}{4}$ **D** none of these

6 The combination table for the set $\{p, q, r, s\}$ under $\oplus$ is given in Fig. 14.8.

$\oplus$	p	q	r	s
p	q	r	p	s
q	r	s	q	p
r	p	q	r	s
s	s	p	s	r

Fig. 14.8

Which of the following statements is false?
   A r is the identity   **B** $\oplus$ is associative
   C the inverse of p is q **D** the table doesn't form a group

7 If $a \circ b$ denotes the remainder when ab is divided by 11, then the solution set of the equation $9 \circ y = 1$ is:
   A $\{5\}$ **B** $\{5, 0\}$ **C** $\{2\}$ **D** none of these

$\oplus$	a	b	c
a	b	a	c
b	a	c	b
c	c	b	a

Fig. 14.9

8 Figure 14.9 shows the composition table for an operation $\oplus$, defined on the set $\{a, b, c\}$. The solution of the equation

$$(b \oplus c) \oplus x = a \qquad\qquad \text{is:}$$

   A a **B** b **C** c **D** none of these

9 The set of functions $f_1 : x \mapsto x$, $f_2 : x \mapsto \dfrac{1}{x}$, $f_3 : x \mapsto -x$, and f_4 form a group under composition of functions. It follows that f_4 is:
   A $x \mapsto 0$ **B** $x \mapsto 1$ **C** $x \mapsto \dfrac{-1}{x}$ **D** none of these

10 The set of all 2 x 2 matrices under matrix multiplication does not form a group because
   A   the set is not closed
   B   multiplication is not always commutative
   C   the inverse does not always exist
   D   multiplication is not always associative
11 Figure 14.10 shows the composition table for an operation * defined on the set {1, 2, 3, 4}.

†	p	q	r	t
p	t	r	p	q
q	r	t	q	p
r	p	q	r	t
t	q	p	t	r

Fig. 14.10

(a) State the identity element.
(b) Solve the equation $(x * 2) * 3 = 1$.

12 The composition table for the set {p, q, r, t} under the operation † is given in Fig. 14.11.

*	1	2	3	4
1	2	4	1	3
2	4	3	2	1
3	1	2	3	4
4	3	1	4	2

Fig. 14.11

(a) State the identity element.
(b) What is the inverse of q?
(c) Is it true that $p † (q † t) = (p † q) † t$?

13 If $a * b = \dfrac{a + b}{b}$, find $3 * (4 * 5)$ and $(3 * 4) * 5$.
   (a) Find a value y such that $y * 2 = 7$.
   (b) Find a value x such that $3 * x = 4$.

14 State whether the following sets are closed under the given operations:
   (a) {even numbers} under addition
   (b) {odd numbers} under addition
   (c) {odd numbers} under multiplication
   (d) {prime numbers} under addition
   (e) {real numbers} under multiplication
   (f) {1, 2, 3, 4} under subtraction

15 Which of the following operations are associative?
   (a) division
   (b) *, where $a * b = a^2 + b^2$
   (c) matrix addition

(d) matrix multiplication

(e) †, where $a \dagger b = \dfrac{a + b}{2}$

(f) set intersection

(g) #, where $a \# b = a + b + ab$

(h) π, where $a \pi b = \dfrac{a + b}{a - b}$

16 Which of the following operations are commutative?
   (a) multiplication of real numbers
   (b) *, where $a * b = a^2 + b^2$
   (c) $\circ$, where $a \circ b = a^2 - b^2$
   (d) multiplication of all 2×2 matrices
   (e) $\oplus$, where $a \oplus b$ denotes the average of a and b
   (f) $, where $a \$ b = \dfrac{a^2 + b^2}{a + b}$

17 The operation * is defined by the table shown in Fig. 14.12 for the set $\{a, b, c, d\}$.

*	a	b	c	d
a	d	c	b	a
b	c	a	d	b
c	b	d	a	c
d	a	b	c	d

Fig. 14.12

   (a) State the identity.
   (b) Find the inverse of c.
   (c) Solve the equation $(a * x) * b = a$

18 If $x \dagger y = x + y + 2$, and $p \circ q = \dfrac{p - 2q - 2}{q + 2}$;

   (a) Show † is associative.
   (b) (i) find $(12 \circ 3) \circ 1$ and $12 \circ (3 \circ 1)$
       (ii) find $(-1 \circ 0)$ and $(0 \circ -1)$
       What do these results suggest?
   (c) What is the identity element for †?

19 The operation * on two numbers means, 'Add the numbers together and divide by 2, ignoring any remainder.'

   For example, $\dfrac{6 + 8}{2} = \dfrac{14}{2} = 7$ and therefore $6 * 8 = 7$.

   Similarly, $\dfrac{6 + 7}{2} = 6$, remainder 1, and therefore $6 * 7 = 6$.

   (a) Write down the values of
       (i) $3 * 7$
       (ii) $2 * 5$
   (b) Find the possible values of x for which $3 * x = 5$
   (c) For the sets $\{1, 2, 3\}$ and $\{2, 4, 6\}$ under the operation *, copy and complete the following tables:

*	1	2	3
1	1		2
2		2	
3	2		3

*	2	4	6
2	2		4
4		4	
6	4		6

Fig. 14.13   **Fig. 14.14**

(d) Use the tables in (c) to enable you to state a property for the integers in a set of integers if the set is to be closed under the operation *.   [UCLES]

20 The whole of this question should be answered on graph paper.

(a) Work out $\begin{pmatrix} 2 & 1 \\ -1 & 0 \end{pmatrix}\begin{pmatrix} 0 & 1 & 0 & -1 \\ 0 & 1 & 2 & 1 \end{pmatrix}$.

(b) Using a scale of 4 cm to 1 unit on each axis, draw on your graph paper the figure $OABC$, where O is the origin and A, B and C have position vectors $\begin{pmatrix} 1 \\ 1 \end{pmatrix}, \begin{pmatrix} 0 \\ 2 \end{pmatrix}$ and $\begin{pmatrix} -1 \\ 1 \end{pmatrix}$ respectively. The figure $OA'B'C'$ is obtained by applying the transformation represented by the matrix $M = \begin{pmatrix} 2 & 1 \\ -1 & 0 \end{pmatrix}$.

(c) State fully the transformation represented by the matrix M.

(d) The square $OFAG$ is such that its vertices O, F, A and G are respectively at the points $(0, 0)$, $(1, 0)$, $(1, 1)$ and $(0, 1)$. Using the same axes, draw the square $OFAG$.

(e) Find the matrix which represents the transformation which maps $OFAG$ onto $OABC$.

(f) Hence, calculate the single matrix which represents the transformation which maps $OA'B'C'$ onto the square $OFAG$.   [UCLES]

21 A transformation P of the plane maps the square A $(2, 2)$, B $(-2, 2)$, C $(-2, -2)$, D $(2, -2)$ onto itself so that A maps onto A, B maps onto D, C maps onto C and D maps onto B. In this case, we can write $P : ABCD \rightarrow ADCB$.

Similarly, transformations of the plane Q, R and S which map $ABCD$ onto itself are defined by:

$Q : ABCD \rightarrow CBAD$,
$R : ABCD \rightarrow CDAB$,
$S : ABCD \rightarrow ABCD$.

(a) Write down the 2×2 matrices **P**, **Q**, **R** and **S** which represent *each* of these transformations. Hence express the matrix product $(\mathbf{QP})\mathbf{R}$ as a 2×2 matrix.

(b) Write down the composition table for the set $\{\mathbf{P, Q, R, S}\}$ under matrix multiplication.

(c) Explain carefully why the set $\{\mathbf{P, Q, R, S}\}$ under matrix multiplication forms a group. (You should demonstrate associativity using only **Q**, **P** and **R** in that order.)   [AEB]

15 The Calculus

15.1 Gradient of a Curve (Differentiation)

Referring to Fig. 15.1, A is the point (x, y) on the curve $y = x^2$, and B is the point $(x + h, y + k)$ on the same curve.

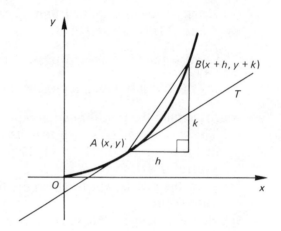

Fig. 15.1

At B, we have

$$y + k = (x + h)^2 = x^2 + 2hx + h^2. \tag{1}$$

(with "equal" bracketing the two underlined terms)

But $y = x^2$,

$$\therefore \ k = 2hx + h^2. \tag{2}$$

Divide (2) by h:

$$\frac{k}{h} = 2x + h.$$

If h becomes very small, the right-hand side gets very near to $2x$.

But $\dfrac{k}{h}$ is the gradient of AB, and as h gets smaller and smaller the gradient of AB is getting nearer to the gradient of the tangent AT at A.

Hence, the gradient of the tangent at $A = 2x$.

The symbol $\dfrac{dy}{dx}$ (pronounced 'd-y by d-x') is used for the gradient.

Hence $\dfrac{dy}{dx} = 2x$.

If we start with $y = x^3$, a similar method gives $\dfrac{dy}{dx} = 3x^2$.

(You should check this.)

246

In general, if $y = x^n$, $\dfrac{dy}{dx} = nx^{n-1}$.

This rule also applies if n is not a positive integer.
It is in fact true for all values of n positive or negative.

The process of finding $\dfrac{dy}{dx}$ is called *differentiation*.

Using function notation, $\dfrac{dy}{dx}$ is denoted by $f'(x)$.

Notes:

(1) If $y = kx^n$, $\dfrac{dy}{dx} = knx^{n-1}$ where k is a constant.

(2) If $y = kx$, $\dfrac{dy}{dx} = kx^0 = k$.

(3) If y is the sum or difference of a number of terms, then $\dfrac{dy}{dx}$ can be found by differentiating each term separately and adding or subtracting the term. See the following example:

Worked Example 15.1

Differentiate the following with respect to x:

(a) $4x^3 + 2x$ (b) $\dfrac{1}{x^3}$ (c) $2\sqrt{x}$ (d) $\dfrac{x^2 + 1}{x^2}$ (e) $2(x + 1)^3$

Solution

(a) $y = 4x^3 + 2x$ $\dfrac{dy}{dx} = 12x^2 + 2$.

(b) $y = \dfrac{1}{x^3}$ must be rewritten as $y = x^{-3}$,

$\therefore \dfrac{dy}{dx} = -3x^{-4} = \dfrac{-3}{x^4}$.

(c) $y = 2\sqrt{x} = 2x^{1/2}$

$\therefore \dfrac{dy}{dx} = 2 \times \tfrac{1}{2}x^{-1/2} = x^{-1/2} = \dfrac{1}{\sqrt{x}}$.

(d) $y = \dfrac{x^2 + 1}{x^2} = \dfrac{x^2}{x^2} + \dfrac{1}{x^2} = 1 + x^{-2}$

$\therefore \dfrac{dy}{dx} = -2x^{-3} = \dfrac{-2}{x^3}$.

(e) $y = 2(x + 1)^3$.

This must be multiplied out, although there are simpler techniques.

$$\therefore \ y = 2(x + 1)(x + 1)^2$$
$$= 2(x + 1)(x^2 + 2x + 1) = 2(x^3 + 2x^2 + x + x^2 + 2x + 1)$$
$$= 2x^3 + 6x^2 + 6x + 2$$
$$\therefore \frac{dy}{dx} = 6x^2 + 12x + 6.$$

15.2 Equation of the Tangent to a Curve

The following example illustrates the use of the calculus in finding the equation of a tangent to a curve.

Worked Example 15.2

Find the equation of the tangent to the curve $y = x^3 - 4x + 1$ at the point P where $x = 2$.

Solution

Always find the coordinates of the point first.

If $x = 2$, $y = 2^3 - 4 \times 2 + 1 = 1$.

$\therefore \ P$ is the point $(2, 1)$.

The equation of a straight line is $y = mx + c$.

To find m, the gradient, differentiate

$$\frac{dy}{dx} = 3x^2 - 4.$$

$\therefore$ If $x = 2$, $m = 3 \times 4 - 4 = 8$.

The equation is $y = 8x + c$.

It has to pass through $(2, 1)$

$\therefore \ 1 = 8 \times 2 + c$, $\therefore \ c = -15$.

The equation of the tangent is $y = 8x - 15$.

15.3 Equation of a Normal to a Curve

A *normal* to a curve is a straight line at right angles to the tangent, passing through the point of contact. If two lines of gradients m_1 and m_2 are at right angles, then it can be shown that $m_1 m_2 = -1$.

To find the equation of the normal to the curve, for Example 15.2, proceed as follows:

Gradient of the tangent = 8.

$\therefore$ Gradient of the normal $= -\frac{1}{8}$. $\quad (8 \times -\frac{1}{8} = -1.)$

$\therefore$ Equation of the normal is $y = -\frac{1}{8}x + c$.

It passes through $(2, 1)$, $\therefore \ 1 = -\frac{1}{8} \times 2 + c$,

$\therefore \ c = \frac{5}{4}$.

The equation is $y = -\frac{1}{8}x + \frac{5}{4}$,

i.e. $\qquad\qquad\qquad\qquad\qquad\qquad 8y = -x + 10.$

15.4 Rates of Change (Velocity and Acceleration)

The calculus is particularly useful in problems related to time.

Figure 15.2 shows a displacement–time graph for an object moving with variable speed. The gradient of the curve at any time gives the velocity; see section 4.12.

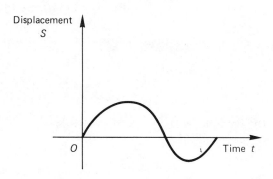

Fig. 15.2

Since we are using (t, s) instead of (x, y), it follows that:

$$\text{velocity} = \frac{ds}{dt}.$$

Figure 15.3 shows a velocity–time curve for a body with non-uniform acceleration. Since the gradient of a velocity–time graph gives acceleration, see section 4.13, we have

$$\text{acceleration} = \frac{dv}{dt}.$$

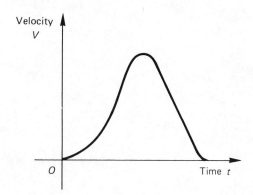

Fig. 15.3

Another way of writing this is $\frac{d^2 s}{dt^2}$.

(This means basically differentiate s twice)

Hence

$$\frac{d^2 s}{dt^2} = \frac{dv}{dt}.$$

Another way of using rates is shown in Worked Example 15.5. Any more detailed work is beyond the scope of this book.

15.5 Integration

If you are given $\dfrac{dy}{dx} = 6x^5$, what was the original function y that was differentiated? The answer given would probably be $y = x^6$. However, it could be $y = x^6 + 8$, $y = x^6 - 15$, i.e. $y = x^6 + c$ where c is an unknown number, called the *constant of integration.* $x^6 + c$ is called the *indefinite integral* of $6x^5$ with respect to x.

The mathematical symbol for integration is $\displaystyle\int \ldots dx$,

$$\therefore \int 6x^5 \; dx = x^6 + c;$$

the dx tells us that x is the variable.

In general,

$$\int kx^n \; dx = \frac{kx^{n+1}}{n+1} + c \qquad \text{where } k \text{ is a constant.}$$

This formula is true for all values of n except -1. To find $\displaystyle\int \frac{1}{x} \; dx$ is beyond the scope of this book.

Worked Example 15.3

Find the integrals of the following functions with respect to x:

(a) $2x^3 + 4x + 1$ (b) $(4x + 1)^2$ (c) $\dfrac{4x + x^3}{\sqrt{x}}$

(d) $4ax^2$ (a is constant).

Solution

(a) $\displaystyle\int 2x^3 + 4x + 1 \, dx = \frac{2x^4}{4} + \frac{4x^2}{2} + \frac{1 \times x^1}{1} + c$

$$= \frac{x^4}{2} + 2x^2 + x + c.$$

(b) $\displaystyle\int (4x + 1)^2 \; dx = \int 16x^2 + 8x + 1 \, dx$

$$= \frac{16x^3}{3} + 4x^2 + x + c.$$

(c) $\displaystyle\int \frac{4x + x^3}{\sqrt{x}} \; dx = \int 4x^{1/2} + x^{5/2} \; dx$

$$= \frac{4x^{3/2}}{\frac{3}{2}} + \frac{x^{7/2}}{\frac{7}{2}} + c$$

$$= \tfrac{8}{3}x^{3/2} + \tfrac{2}{7}x^{7/2} + c.$$

(d) $\int 4ax^2\ dx$

(the dx tells us that a is considered only as an ordinary number)

$$= \frac{4ax^3}{3} + c.$$

15.6 Definite Integrals

$\displaystyle\int_2^3 4x^2 + 3x + 1\ dx$ is called a *definite integral**.

The number 2 is called the *lower limit*, and 3 the *upper limit* of the integral. It is evaluated as follows:

$$\int_2^3 4x^2 + 3x + 1\,dx = \left[\frac{4x^3}{3} + \frac{3x^2}{2} + x\right]_2^3$$

— note that $+c$ is not required

$$= \underbrace{\left[\frac{4 \times 3^3}{3} + \frac{3 \times 3^2}{2} + 3\right]}_{\substack{\text{substitute upper} \\ \text{limit for } x}} - \underbrace{\left[\frac{4 \times 2^3}{3} + \frac{3 \times 2^2}{2} + 2\right]}_{\substack{\text{substitute lower} \\ \text{limit for } x}}$$

$$= 52\tfrac{1}{2} - 18\tfrac{2}{3} = 33\tfrac{5}{6}.$$

15.7 Area Under a Curve

In section 4.8 we saw how the trapezium rule could be used to approximate areas under curves. The definite integral can be used to find the exact area, providing one or two points are watched carefully.

In Fig. 15.4, to find the shaded area under the curve $y = x^3 + 1$, between $x = -1$

and $x = 2$, find $\displaystyle\int_{-1}^2 x^3 + 1\ dx$

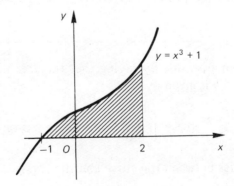

Fig. 15.4

*Note: Large square brackets [ ] are always used to enclose a definite integral once integration has been carried out.

$$= \left[\frac{x^4}{4} + x\right]_{-1}^{2} = 6 - -\tfrac{3}{4} = 6\tfrac{3}{4} \text{ square units.}$$

If part of the curve lies below the axis, care must be taken to evaluate each part of the area lying above and below the axis separately.

To find the area shaded in Fig. 15.5, the following procedure must be adopted:

$$A_1 = \int_{-1}^{0} x^3 - x^2 - 2x \, dx = \left[\frac{x^4}{4} - \frac{x^3}{3} - x^2\right]_{-1}^{0}$$

$$= [0] - [\tfrac{1}{4} + \tfrac{1}{3} - 1] = \tfrac{5}{12} \text{ square units,}$$

$$A_2 = \int_{0}^{2} x^3 - x^2 - 2x \, dx = \left[\frac{x^4}{4} - \frac{x^3}{3} - x^2\right]_{0}^{2}$$

$$= [\tfrac{16}{4} - \tfrac{8}{3} - 4] - [0] = -\tfrac{8}{3} \text{ square units.}$$

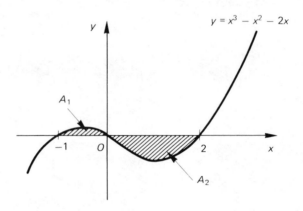

Fig. 15.5

Note that A_2 is negative. This just means that the area lies *below* the x-axis.

The total area $= \tfrac{5}{12} + \tfrac{8}{3} = 3\tfrac{1}{12}$ square units.

It is totally wrong to find $\displaystyle\int_{-1}^{2} x^3 - x^2 - 2x \, dx$;

this would have a given value of $-2\tfrac{1}{4}$ square units.

We can summarise this by saying that the area under the curve $y = f(x)$ between $x = a$ and $x = b$ is given by

$$\text{area} = \int_{a}^{b} y \, dx \qquad \text{where } y \text{ is replaced by } f(x).$$

If the area lies between the curve and the y axis,

$$\text{area} = \int_{c}^{d} x \, dy \qquad \text{where } x \text{ is replaced by } f(y).$$

15.8   Volume of Revolution

If the shaded areas in the previous section are rotated by 360° about the x-axis, a solid will be generated. Its volume can be found by using the formula:

$$\text{Volume} = \pi \int_a^b y^2 \, dx.$$

It is not necessary to worry about whether the area is above the axis below. Hence, if the total area in Fig. 15.5 is rotated by 360° about the x-axis, then

$$\text{Volume} = \pi \int_{-1}^{2} (x^3 - x^2 - 2x)^2 \, dx.$$

Now $(x^3 - x^2 - 2x)(x^3 - x^2 - 2x)$
$$= x^6 - x^5 - 2x^4 - x^5 + x^4 + 2x^3 - 2x^4 + 2x^3 + 4x^2$$
$$= x^6 - 2x^5 - 3x^4 + 4x^3 + 4x^2.$$

Hence,
$$\text{Volume} = \pi \int_{-1}^{2} x^6 - 2x^5 - 3x^4 + 4x^3 + 4x^2 \, dx$$

$$= 14.54 \text{ cubic units.}$$

Check 32

1 Differentiate the following functions with respect to x:

   (a) $3x^6$ (b) $x^3 + 2x^2 - 5$ (c) $\dfrac{4}{x^3}$ (d) $2x^5 - 3$

   (e) $3x + \dfrac{2}{x}$ (f) $2\sqrt{x}$ (g) $3x^{0.2}$ (h) $x^{1/3} + x^{3/4}$

   (i) $\dfrac{1}{3\sqrt{x}}$ (j) $x(x-2)$ (k) $(x+2)(x-1)$ (l) $(x^3+1)^2$

   (m) $\dfrac{x+1}{x}$ (n) $\dfrac{(x+1)^2}{x}$ (o) $\left(x + \dfrac{1}{x}\right)(x-2)$

2 Integrate the following functions with respect to x:

   (a) $x^2 + 1$ (b) $3x^2 - 5x + 1$ (c) $\sqrt{x}$ (d) $\dfrac{x+1}{x^3}$

   (e) $(x-1)^2$ (f) $(x+2)\left(x + \dfrac{2}{x^3}\right)$ (g) $x^3 - 3x + \dfrac{1}{2x^2}$

   (h) $(x^3 - 1)(x^3 + 1)$

3 Find the following definite integrals:

   (a) $\displaystyle\int_1^2 x^2 + 3 \, dx$ (b) $\displaystyle\int_0^1 x^4 - 5 \, dx$ (c) $\displaystyle\int_{-1}^1 x^2 + 2x + 2 \, dx$

   (d) $\displaystyle\int_{-1}^3 x + \dfrac{2}{x^2} \, dx$ (e) $\displaystyle\int_1^3 2x(x+3) \, dx$ (f) $\displaystyle\int_{-2}^0 x(x-2) \, dx$

   (g) $\displaystyle\int_1^4 (x^2 + 1)^2 \, dx$ (h) $\displaystyle\int_1^2 \dfrac{x^2 + 1}{x^2} \, dx$

15.9   Maxima and Minima

The turning points on a graph are called maximum points or minimum points; see Fig. 15.6. At these points, the curve is parallel to the x-axis,

hence
$$\frac{dy}{dx} = 0$$

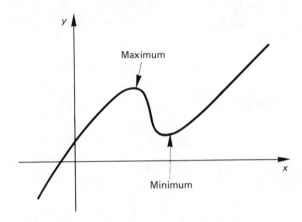

Fig. 15.6

Look at the gradient on each side of the point to find whether it is a maximum or a minimum. See Worked Example 15.6.

Worked Example 15.4

The curve $y = 4x + x^2$ cuts the x-axis at the origin, and point P.
(a) Calculate the coordinates of P, and sketch the graph.
(b) Calculate the coordinates of the point Q on the graph which has a gradient of 2.
(c) Calculate the area between the curve and the x-axis.

Solution

(a) The curve cuts the x-axis when $y = 0$.

Hence, $0 = 4x + x^2 \Rightarrow 0 = x(4 + x)$,
$$\therefore x = 0 \text{ or } -4.$$

The required value is $x = -4$, hence P is the point $(-4, 0)$.
The equation $y = 4x + x^2$ is a parabola, and since the coefficient of x^2 is positive, it is U-shaped. Figure 15.7 shows the curve.
(b) The gradient at any point is found by differentiating;

$$\therefore \frac{dy}{dx} = 4 + 2x.$$

If the gradient is 2, $2 = 4 + 2x$, hence $x = -1$
$\therefore Q$ is $(-1, -3)$.

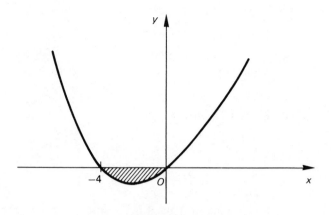

Fig. 15.7

(c) The area is given by the definite integral of the function from $x = -4$ to $x = 0$;

hence

$$\text{area} = \int_{-4}^{0} 4x + x^2 \, dx$$

$$= \left[2x^2 + \frac{x^3}{3} \right]_{-4}^{0} = [0] - [32 - \tfrac{64}{3}] = -10\tfrac{2}{3}.$$

The *negative* sign just indicates that the area is below the x-axis.

Worked Example 15.5

O and P are points on a line. A particle moves along the line in such a way that, t seconds after it is at O, its velocity is v cm/s, where $v = kt - t^2$ and k is a constant. At the time when $t = 6$, the particle is momentarily at rest at P. Find:
(a) the value of k;
(b) the distance OP;
(c) the average speed of the particle between O and P;
(d) the acceleration of the particle when it is at P;
(e) when the particle has zero acceleration. [L]

Solution

If velocity is given and the question requires information about distance and acceleration, integration and differentiation will be required; write

$$v = \frac{ds}{dt}.$$

$$\therefore \frac{ds}{dt} = kt - t^2. \tag{1}$$

Integrate with respect to t:

$$s = k\frac{t^2}{2} - \frac{t^3}{3} + C. \qquad \text{(Do not forget the } +C.)$$

Since $s = 0$ when $t = 0$, it follows that $C = 0$.

$$\therefore s = k\frac{t^2}{2} - \frac{t^3}{3}. \tag{2}$$

Also, $\dfrac{d^2 s}{dt^2} = k - 2t.$ (3)

(a) Since $v = 0$ when $t = 6$,
using (1), $\qquad\qquad 0 = 6k - 36$
$\qquad\qquad\qquad\qquad \therefore\ k = 6.$

(b) Using (2), with $t = 6$,

$$s = \dfrac{k \times 6^2}{2} - \dfrac{6^3}{3} = \dfrac{6 \times 6^2}{2} - \dfrac{6^3}{3}$$
$$= 36.$$

$\therefore OP = 36$ cm.

(c) Do not be confused here, since average speed does not involve differentiation.

Average speed $= \dfrac{OP}{6} = 6$ cm/s.

(d) Using (3) when $t = 6$, ·

$$\dfrac{d^2 s}{dt^2} = 6 - 2 \times 6 = -6,$$

$\therefore$ acceleration $= -6$ cm/s^2 (particle is slowing down).

(e) Using (3) with $\dfrac{d^2 s}{dt^2} = 0,$

$$0 = 6 - 2t$$
$$\therefore\ t = 3.$$

Worked Example 15.6

Find the coordinates of the turning points on the curve $y = \dfrac{1}{x} + x$, and distinguish between them.

Solution

$$\text{If } y = \dfrac{1}{x} + x,$$

$$\dfrac{dy}{dx} = -\dfrac{1}{x^2} + 1 = 0 \qquad \text{if } 1 = \dfrac{1}{x^2},$$

$$\therefore\ x^2 = 1,\ x = \pm 1.$$

Consider $x = 1$. $\qquad$ If $x = 0.8$, $\qquad \dfrac{dy}{dx} = -\dfrac{1}{0.8^2} + 1 = -0.6.$

$\qquad\qquad\qquad$ If $x = 1.2$, $\qquad \dfrac{dy}{dx} = -\dfrac{1}{1.2^2} + 1 = 0.3.$

The gradient goes from negative to positive as x increases. Hence $x = 1$ is a minimum.

Consider $x = -1$. $\qquad$ If $x = -1.2$, $\qquad \dfrac{dy}{dx} = 0.3.$

$\qquad\qquad\qquad$ If $x = -0.8$, $\qquad \dfrac{dy}{dy} = -0.6.$

The gradient goes from positive to negative as x increases. Hence $x = -1$ is a maximum.

$$\therefore x = 1, y = 2 \text{ is a minimum,}$$
$$x = -1, y = -2 \text{ is a maximum.}$$

Exercise 15

1 If $y = x + \dfrac{1}{x}$, then the gradient of the tangent at the point $(1, 2)$ is:

    A 0   **B** 2   **C** -1 **D** 1

2 The area of the region bounded by the curve $y = \dfrac{1}{x^2}$, and the straight lines $x = 1$, $x = 2$ and $y = 0$ is:

    A 1   **B** $\frac{1}{2}$ **C** 2   **D** $1\frac{3}{4}$

3 If $\displaystyle\int_0^2 f(x)\,dx = 4$ and $\displaystyle\int_0^2 g(x)\,dx = 3$, then $\displaystyle\int_0^2 2f(x) + 3g(x) + 4\,dx$ is:

    A 9   **B** 21   **C** 25   **D** none of these

4 A circular oil patch is increasing at a rate of 8 m² per second. When the radius of the circle is 12 m, its radius is increasing at a rate of:

    A 3π m/s   **B** 3 m/s   **C** $\frac{1}{3}$ m/s   **D** $\dfrac{1}{3\pi}$ m/s

5 The speed v m/s of an object moving in a straight line AB, t seconds after it leaves A, is given by $v = 8t - 32t^3$. The distance it has travelled when it first comes to rest is:

    A 1 m   **B** $\frac{1}{2}$ m   **C** $1\frac{1}{2}$ m   **D** none of these

6 The minimum value of $4(x - 3)^2 - 5$, $x \in R$ is:

    A 29   **B** 19   **C** -3 **D** -5 **E** -17 [SEB]

7 If the function $y = 2x^2 + px + 7$ has a minimum value where $x = 2.5$, then the value of p is:

    A -10 **B** -5 **C** 0   **D** 5   **E** 10   [LOND]

8 The shaded area in Fig. 15.8 is rotated by 360° about the y-axis. The volume formed is:

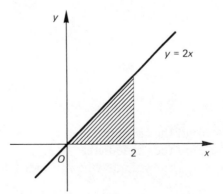

Fig. 15.8

    A $\dfrac{16\pi}{3}$ **B** $\dfrac{32\pi}{3}$ **C** 16π **D** none of these

9 The range of values of x for which the gradient of the curve $y = x^3 - 3x$ is increasing is:

    A $x > 1$ **B** $\{x < -1\} \cup \{x > 1\}$ **C** $x > -1$ **D** $x > 0$

10 The equation of the tangent to the curve $y = x^3 - 4x$ at the point where $x = \dfrac{-2}{\sqrt{3}}$ is:

A $y = x + 1$ B $y = 0$ C $y = \dfrac{16}{3\sqrt{3}}$ D none of these

11 At any point on a curve, the gradient is given by $\dfrac{dy}{dx} = 6x + 1$. If the curve passes through the point $(1, 0)$, find the equation of the curve.
   (a) Find the coordinates of the turning point on the curve, stating whether it is a maximum or a minimum.
   (b) Find the area between the curve and the x-axis.
   (c) What is the equation of the straight line that divides this area in half?

12 The velocity v m/s of a point P, moving along a straight line passing through a point O, is given in terms of the time t seconds by

$$v = -6t^2 + 30t - 36.$$

The point P is at O when $t = 0$.
Determine expressions in terms of t for
(a) the distance OP; (b) the acceleration of P.
   Show that P is at rest twice and find the distance between these rest positions. Find also the maximum velocity of P.   [L]

13 The diagram (Fig. 15.9) shows the area between the curves $y = x^2$, $y = 8 - x^2$, and the y-axis.

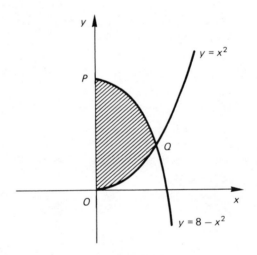

Fig. 15.9

(a) Find the coordinates of Q.
(b) Find the shaded area.
(c) Find the volume when this area is rotated about the y-axis.

14 Find: (a) $\displaystyle\int_0^4 (2x + 1)^2 \, dx$ (b) $\displaystyle\int_1^2 \dfrac{(x + 1)^2}{x^4} \, dx$ (c) $\displaystyle\int_0^1 1 - (x + 1)^2 \, dx$

15 A rectangular cuboid is made from a thin metal sheet. Its measurements are $4x$ cm by x cm by y cm. Find an expression for the total surface area S of the cuboid, in terms of x and y.
   The cuboid is made to have a surface area of 100 cm². If V cm³ is the volume of the cuboid, show that $V = \frac{2}{5}(100x - 8x^3)$. Hence show that the maximum value of V is approximately 54 cm³.

16 A stone is thrown from the top of a tower of height 100 m. After t seconds, its distance from the ground H cm is given by

$$H = 5(20 + t - t^2).$$

(a) What height does the stone reach above the tower before falling?
(b) How long does the stone take to reach the ground?
(c) With what speed does the stone hit the ground?

17 A rectangular block of metal has a square base. The sides of the base are each x cm long and the height of the block is y cm. The sum of the lengths of the twelve edges of the block is 132 cm.

(a) Write down an expression in terms of x and y for the surface area, A cm^2, of the block.
(b) Obtain a formula for y in terms of x, and hence prove that $A = 132\,x - 6x^2$.
(c) Prove that as x varies, the greatest possible value of A is 726, and state the dimensions of the block in this case. [AEB]

18 (a) Calculate the gradient of the tangent at $Q\,(-1, 5)$ on the curve $y = 4x^2 + 1$, and hence find the equation of the tangent at Q.
(b) On the curve $y = 2x^3 - 21x^2 + 36x$, calculate the maximum and minimum points, carefully distinguishing between them.

19 (a) (i) If $y = x^2 - x - 5$, write down the expression for $\dfrac{dy}{dx}$ and hence find the values of x for which $y = \dfrac{dy}{dx}$.

(ii) Evaluate $\displaystyle\int_{-1}^{2} (4 - x^2)\,dx$ and show clearly in a neat sketch the area represented by this integral.

(b) A lift moves from rest at ground level to rest at the top of a building. When it has been moving for t minutes its height, h metres, above the ground is given by $h = 20(t^2 - \frac{1}{3}t^3)$.
(i) Obtain expressions for its velocity and acceleration in terms of t.
(ii) Calculate the value of t when the velocity is a maximum.
(iii) Find the value of t when the lift comes to rest at the top of the building and hence calculate the height of the building. [NISEC]

20 Particle P moves along a straight line AX, 50 cm long. At time $t = 0$, P is at A and t seconds later its velocity v cm/s is given by $v = 15 + 4t - 3t^2$.
(a) Write down expressions for
(i) the acceleration of P at time t seconds, and
(ii) the distance of P from A at time t seconds.
(b) (i) Find when P is instantaneously at rest.
(ii) How far is P from X at this time?
(c) Find the period of time during which the acceleration of P is positive.
(d) Show that P returns to A after 5 seconds.
(e) How far does P travel in the fifth second? [NISEC]

21 A particle moves in a straight line so that its distance, s cm, from a fixed point O is given after time t seconds by the formula $s = 4t^3 - 7t^2 - 6t$. Calculate:
(a) the distance from O after 3 seconds;
(b) the velocity after 2 seconds;
(c) the acceleration when it is at O;
(d) the time when the particle is at rest;
(e) the distance the particle travels during the third second of its motion.

22 A particle moves in a straight line such that its velocity v m/s is given after time t seconds by the equation $v = 32 + 4t - t^2$.

Calculate:

(a) its initial velocity;

(b) the acceleration when it comes to rest;

(c) the distance travelled in the seventh second.

23 (a) Integrate $3.2x^7 - \dfrac{2}{x^2}$ with respect to x.

(b) If $\displaystyle\int_1^a 3(x + 1)^2 \, \mathrm{d}x = a^3 + 11$, find the possible values of a.

24 (a) The equation of a curve is $y = x^3 + 2x^2 + cx + 36$ where c is a constant.

(i) Write down the expression for $\dfrac{\mathrm{d}y}{\mathrm{d}x}$;

(ii) P is a turning point on the curve. Given that the x coordinate of P is $\frac{5}{3}$, find the value of c;

(iii) Find the coordinates of the second turning point Q on this curve;

(iv) Determine which of the points P, Q is a maximum and which is a minimum turning point.

(b) Figure 15.10 shows part of the curve whose equation is $y = x^2$. The line $y = 4$ cuts the curve at the points A and B. Find the area enclosed between the chord AB and the curve. [NISEC]

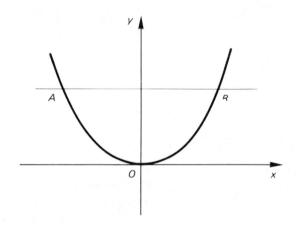

Fig. 15.10

25 The curve $y = 15x - 3x^2$ meets the x-axis at the points O and P, where O is the origin. The curve meets the line $y = 3x$ at the points O and Q.

(a) Calculate the x-coordinate of the point P.

(b) Use the methods of the calculus to find the gradient of the curve at P.

(c) Calculate the coordinates of the point Q.

(d) Use the methods of the calculus to find the area of the region bounded by the line segment OP, the line segment OQ and the arc PQ of the curve.

[JMB]

26 (a) Find the coordinates of the points of intersection of the straight line whose Cartesian equation is $y = x$ and the curve whose Cartesian equation is $y = 3x - x^2$. The function f is defined for all real numbers x satisfying $0 \leqslant x \leqslant 3$ by

$$\text{f} : x \mapsto \begin{cases} x & \text{for } 0 \leqslant x \leqslant 2 \\ 3x - x^2 & \text{for } 2 < x \leqslant 3. \end{cases}$$

(b) Sketch the graph of this function.

(c) Calculate the area enclosed by this graph and the x-axis.

(d) Given that the line $x = C$ divides this region into two regions of equal area, find the value of C correct to two decimal places.

27 The curve whose equation is $y = 2x - x^2$ intersects the x-axis at the origin and again at P.

(a) Calculate the x-coordinate of P;

(b) Sketch the graph of $y = 2x - x^2$;

(c) Calculate the coordinates of the point on the curve at which the gradient equals 1;

(d) The area bounded by the curve and the x-axis is rotated completely about the x-axis. Calculate the volume of the solid of revolution generated.   [L]

28 (a) Sketch the curve whose equation is

$y = x^2 - 6x$.

(b) Calculate the gradient of the curve at the point P (5, −5).

(c) Find the coordinates of the point V on the curve at which the tangent is parallel to the x-axis.

(d) Given that the point O is the origin and R is the point on the curve whose coordinates are (8, 16), show that $ORPV$ is a trapezium.

(e) The point N is the foot of the perpendicular from R to the x-axis. Calculate the area of the region enclosed by NR, the x-axis and that part of the curve which is above the x-axis.   [JMB]

Answers

Check 1

1 (a) {11, 13, 17, 19, 23, 29}   (b) {$1\frac{1}{2}$}   (c) ϕ
   (d) {(0, 8), (1, 7), . . ., (8, 0)}   (e) {30, 150}

2 (a) {4, 5}   (b) {3, 5, 7, 8}   (c) {1, 4, 6}
   (d) {2, 3, 4, 5, 6, 8}   (e) {1, 2, 4, 6, 7}   (f) { 3, 5, 8 }
   (g) {1, 4, 5, 6, 7}

4 8, 13

5 19

6 (a) {3, 4, 5, 6, 7, 8}   (b) {4, 5, 6}   (c) {$-\infty$, . . ., 0, 1, 2}

Check 2

1 8.69, 26.13, 0.09, 0.10, 2.02, 2.19, 10.00

2 86.4, 187, 0.0900, 2.00, 170, 3.00, 8700, 10 000

Check 3

(a) 94   (b) 140   (c) 1128   (d) 11   (e) 61   (f) 35   (g) 420
(h) 587   (i) 2133   (j) 2343

Check 4

1 110000, 1010110, 10001, 10111011000, 10111, 11001111, 111110, 1011100

2 13, 27, 333, 125

Check 5

1 10100   **2** 100001   **3** 1111   **4** 1443   **5** 10100   **6** 35552
7 417   **8** 31   **9** 33   **10** −111   **11** 11110   **12** 14
13 1202   **14** 2225   **15** 1215

Check 6

1 $1\frac{7}{20}$ **2** $\frac{8}{15}$ **3** $1\frac{1}{12}$ **4** $\frac{3}{20}$ **5** $\frac{1}{10}$ **6** $\frac{5}{24}$
7 $\frac{1}{4}$ **8** $1\frac{1}{8}$ **9** $1\frac{1}{2}$ **10** $\frac{3}{5}$ **11** $6\frac{5}{6}$ **12** $1\frac{9}{16}$
13 $\frac{1}{5}$ **14** $2\frac{4}{7}$ **15** $\frac{31}{240}$ **16** 3   **17** $29\frac{1}{2}$ **18** $1\frac{11}{25}$
19 $8\frac{8}{9}$ **20** $\frac{1}{3}$

Check 7

1 0.6	2 0.12	3 0.625	4 0.714	5 0.1̇8̇	6 0.52
7 0.27̇	8 0.368	9 0.4375	10 0.354		

Check 8

1 40, 279, $16\frac{2}{3}$, 3.75, 0.5, 55, 150, 9.8

2 $\frac{1}{3}$, $\frac{18}{25}$, $\frac{1}{8}$, $\frac{3}{500}$, $\frac{3}{25}$, $\frac{73}{200}$

Check 9

1 4.86×10^2, 6.93×10^3, 8.57×10^5, 10^{-2}, 9.6×10^{-2}, 8.31×10^{-3}, 9.09×10^{-2}

2 45 000, 3840, 8 010 000, 0.085, 0.000 695, 0.000 008 801

Check 10

9, 0.007, 40, 0.1, 1, 100, 400 000, 0.006

Check 11

1 $6\sqrt{2}$, $7\sqrt{6}$, $\sqrt{2}$, $5\sqrt{10}$, 0, $\sqrt{7}$, $-\sqrt{11}$, $6\sqrt{7}$, $10\sqrt{2} - 2\sqrt{3}$, $5\sqrt{5}$

2 $\frac{\sqrt{2}}{2}$, $\frac{\sqrt{5}}{25}$, $\frac{\sqrt{2}}{2}$, $\frac{\sqrt{2}}{2}$, $\frac{\sqrt{10}}{2}$, $\frac{\sqrt{5}}{5}$, $\frac{\sqrt{7}+\sqrt{2}}{5}$

Check 12

1 -6	2 -16	3 12	4 $-1\frac{1}{8}$	5 $-\frac{2}{3}$	6 -7	7 $-\frac{3}{20}$
8 $-\frac{1}{32}$	9 $-1\frac{57}{160}$	10 $1\frac{3}{32}$	11 -10	12 $7\frac{14}{15}$	13 $\frac{9}{40}$	14 0
15 $35\frac{47}{80}$						

Check 13

1 100	2 9	3 0.3	4 $\frac{1}{25}$	5 $\frac{1}{3}$	6 4	7 1
8 7	9 4	10 $\frac{1}{25}$	11 $2x^3$	12 $5p^{-8}/4$		13 $12x^{-6}$
14 $-27m^3$		15 $3x^2$	16 $5y^8$	17 $4q^{4/3}$		18 u
19 $9v^4$		20 $\frac{8}{27}$				

Check 14

1 $y(a + 4y)$	2 $t(3t^2 + 2t + 5)$	3 $p^2q^2(q - p)$	4 $4(p - 2)$
5 $(p + q)(y - z)$	6 $(r - p)(q + s)$	7 $(t - 3)(t - p)$	
8 $(x - y)(a + b + c)$	9 $(x - 3)(x - 8)$	10 $(x + 3)(x - 1)$	
11 $(x + 13)(x - 2)$	12 impossible	13 $(2x + 3)(x + 5)$	
14 $(2x + 1)(2x + 3)$	15 $(3x - 1)(2x - 1)$	16 $(9x + 2)(x - 1)$	
17 $(4x + 1)(x - 1)$	18 $(2x - 5)(2x + 5)$	19 $(1 - x)(1 + x)$	

20 $\pi(R - r)(R + r)$ **21** $x(x - 6)$ **22** $(11xy - 2)(11xy + 2)$
23 $(2x - 3y)^2$ **24** $(5x - y)^2$ **25** $(5 - b)^2$

Check 15

1 (a) $\dfrac{x + 1}{x^2}$ (b) $\dfrac{2b + 3}{ab}$ (c) $\dfrac{8t - 3}{2t^2}$ (d) $\dfrac{13x}{12}$ (e) $\dfrac{25}{3y}$

(f) $\dfrac{28x^2 - 15y^2}{21xy}$ (g) $\dfrac{4x^2 + 1}{x}$ (h) $\dfrac{7a + 12b}{8}$ (i) $\dfrac{2x - 1}{6}$

(j) $\dfrac{7x + 5}{(x - 1)(x + 2)}$ (k) $\dfrac{-1}{x(x - 1)}$ (l) $\dfrac{x^2 - 2x + 1}{(x - 3)(x - 2)}$ (m) $\dfrac{4x}{x^2 - 1}$

2 (a) $\dfrac{2}{y}$ (b) $\dfrac{4tq}{3}$ (c) $\dfrac{3q}{5t}$ (d) $\dfrac{8x}{15y}$ (e) $\dfrac{8}{15}$

(f) $\dfrac{2q}{3}$ (g) $\dfrac{c}{a + b}$ (h) $\dfrac{25}{18t}$ (i) $4tp$ (j) $\dfrac{a + b}{a - b}$

Check 16

1 $\dfrac{2x - 5}{3}$ **2** $kq + \dfrac{py}{2000}$ **3** $\dfrac{xy}{10}$ **4** $150 + x + y$

5 $yt + v(50 - y)$ **6** $\dfrac{10p}{11}$ **7** $4f + 6s + 9$ **8** 1.2%

9 $\dfrac{(E - D)I}{100}$ **10** $85W + M$

Check 17

1 (a) 2   (b) 4   (c) $-\frac{4}{3}$ (d) $-\frac{10}{3}$ (e) $\frac{19}{5}$ (f) $\frac{8}{5}$
(g) $1\frac{1}{2}$ (h) $-\frac{8}{5}$ (i) $\frac{15}{26}$ (j) $\frac{1}{2}$ (k) $\frac{34}{23}$ (l) -9
(m) $\frac{7}{9}$ (n) 8   (o) $-\frac{13}{5}$
2 (a) $3, -1$ (b) $2, 8$ (c) $\frac{1}{2}, -1$ (d) $\frac{1}{3}, 3$ (e) 1   (f) $0, 9$
(g) $1.85, -4.85$ (h) $-1 \pm \sqrt{2}$ (i) $-4 \pm \sqrt{19}$ (j) $-0.11, 2.36$
(k) $1.82, 0.18$ (l) $-6 \pm \sqrt{19}$ (m) no solution
(n) $\pm\dfrac{2}{\sqrt{3}}$ (o) 1
3 (a) $5, 2$ (b) $-2, -5$ (c) $2, 1$ (d) $2, 4$ (e) $\frac{11}{10}, \frac{3}{10}$ (f) $3, 2$
(g) $1\frac{1}{2}, -3$ (h) $6, -\frac{1}{2}$ (i) $5, 1$ (j) $-3, -\frac{11}{3}$

Check 18

1 (a) $(x - 1)(x^2 + x + 1)$ (b) $(x + 2)(x - 1)(x + 1)$
(c) $(x - 2)(x^2 + 9x + 15)$ (d) $(x - 1)(x - 2)(x - 3)$
(e) $(x - 1)(x + 1)(2x - 1)$ (f) $(3x - 1)(x^2 + 3x + 1)$
(g) $(x + 2)(x - 3)(x - 5)$
2 $\frac{9}{8}$ **3** $\frac{4}{9}$ **4** $121\frac{1}{2}$ **5** $-\frac{3}{2}, \frac{1}{2}$

Check 19

1 430, 68.4, 40 000, 5, 0.25, 8.63 × 10⁷, 4000, 125.5
2 6.25 × 10⁻³, 0.08, 9.4 × 10⁻³, 0.895, 8.69 × 10⁻⁴
3 8000, 40, 0.685, 65 000, 1.8 × 10⁻³, 9 × 10⁵
4 9.6 × 10⁴, 28.75, 4.8 × 10⁸
5 0.6, 2.8 × 10⁴, 2 × 10¹²
6 72, 6.94, 8

Check 20

1 10.3, 13.44, 11.72, 15.32, 13.2, 53.92, 92.52, 28.56
2 2.176×10^7
3 93.5 cm²
4 3.5
5 616, 484, 14, $201\frac{1}{7}$, 2.2, 10.68, 1232, 2, 38 808, 5544, $50\frac{2}{7}$, 2, 91 989, 28,
   11.5, 24.64, 91.99, 2.8, $205\frac{1}{3}$, 177.4, 331.4, 8.06, $314\frac{2}{7}$, 204.3, 282.9, 5, 167.2,
   321.2, 7, 7.6, 1797, 905.1, 26.8, 28, 2932, 880, 14.28, 20.

Check 21

1 3.34
2 7.3
3 $\dfrac{\pi}{4}, \dfrac{\pi}{3}, \pi, \dfrac{3\pi}{4}, \dfrac{2\pi}{15}, \dfrac{\pi}{24}$
4 17.2, 171.9, 49.3, 22.5, 120, 179.9
5 8.55 cm, 14 453 km, 0.448 m, 2513 km
6 14.32 cm, 32.17 cm, 630.2 km, 203.7 km
7 86, 57, 5, 1

Check 22

1 $-7 \leqslant y \leqslant 5, \{0, 1, 4, 9\}, \{x \in R, x \neq 0\}, -1 \leqslant y \leqslant 1$
2 $2x^2 + 1, (2x - 3)^2, 4x - 9, 8x^2 - 24x + 19$
3 1, 0, $\frac{3}{4}$
4 3, 11
5 30, 55
8 -4
9 $3y = 2x + 12, y = 3x - 2, 2y = 3x - 6, y + 3x = 10$
10 $(1\frac{1}{2}, 5), (\frac{1}{2}, -\frac{1}{2}), (1, -1\frac{1}{2}), (2\frac{1}{2}, 2\frac{1}{2})$

Check 23

1 $x < \frac{11}{3}$
2 $x < 1$
3 $\{1, 2, \ldots, 10\}$
4 -1
5 $-2 < x < 12$
6 7, 8
7 $-3 < x < 4$

8 $-4.6 < x < -2.2$
9 $\{2, 3, \ldots, 7\}$
10 $-1 < x < 4$
11 $2 < x < 3$
12 $-\sqrt{15} < x < \sqrt{15}$
13 $-3 < x < 2$
14 $0 < x < 2\frac{1}{2}$
15 $x > 2$ or $x < -2$

Check 24

1 $\begin{pmatrix} 1 \\ 2 \end{pmatrix}, \begin{pmatrix} -6 \\ 3 \end{pmatrix}, \begin{pmatrix} 8 \\ 1 \end{pmatrix}, \begin{pmatrix} 5 \\ 0 \end{pmatrix}, \begin{pmatrix} 7 \\ 4 \end{pmatrix}, \sqrt{10}, \sqrt{5}, \sqrt{10} + \sqrt{5}$

2 $b + c, a + b + c, a + b$

3 $b - a, -b - a, 2b$

4 $\begin{pmatrix} -6 \\ -2 \end{pmatrix}, \begin{pmatrix} 7 \\ -2 \end{pmatrix}, \begin{pmatrix} 1 \\ -4 \end{pmatrix}$

5 $(5, -3), \begin{pmatrix} 2 \\ -4 \end{pmatrix}$

6 $(6, 4), (18, 8), \sqrt{265}$

Check 25

1 $\frac{1}{2} \begin{pmatrix} 3 & -2 \\ -5 & 4 \end{pmatrix}$

2 $\begin{pmatrix} 6 & 7 \\ 5 & 6 \end{pmatrix}$

3 impossible

4 $\frac{1}{36} \begin{pmatrix} 9 & -3 \\ 6 & 2 \end{pmatrix}$

5 $\begin{pmatrix} 1 & 0 \\ 0 & -1 \end{pmatrix}$

6 $\frac{1}{78} \begin{pmatrix} 11 & 4 \\ -3 & 6 \end{pmatrix}$

Check 26

1 (a) $\begin{pmatrix} 11 & 5 & -1 \\ -4 & -1 & 4 \end{pmatrix}$ (b) $\begin{pmatrix} -2 & -3 \\ 14 & -6 \\ 7 & -19 \end{pmatrix}$ (c) $\begin{pmatrix} -5 & 10 \\ -13 & 0 \end{pmatrix}$

(d) impossible (e) $\begin{pmatrix} 4 & -4 & -8 \\ 0 & 4 & 12 \\ 16 & 4 & 24 \end{pmatrix}$ (f) impossible

2 $\begin{pmatrix} \frac{7}{4} & \frac{3}{4} \\ -1 & \frac{9}{4} \end{pmatrix}$

$3 \begin{pmatrix} \frac{2}{5} & 0 \\ \frac{1}{5} & -\frac{3}{5} \end{pmatrix}$

4 (a) (10) (b) $\begin{pmatrix} 6 & 3 \\ 2 & 1 \end{pmatrix}$ (c) (-9) (d) impossible

(e) $\begin{pmatrix} 9 & 5 \\ 5 & 2 \end{pmatrix}$ (f) $\begin{pmatrix} -4 \\ 5 \end{pmatrix}$ (g) $\begin{pmatrix} 3 \\ 3 \end{pmatrix}$

(h) $\begin{pmatrix} 6 & 9 & 0 \\ 3 & 6 & 3 \end{pmatrix}$ (i) $\begin{pmatrix} 2 & 10 & 2 \\ 13 & 11 & 31 \\ 17 & 2 & 38 \end{pmatrix}$

$5 \begin{pmatrix} -4 & 0 \\ -9 & -3 \end{pmatrix}$

6 (a) $\frac{25}{11}, \frac{-16}{11}$ (b) $\frac{62}{37}, \frac{-17}{37}$ (c) $\frac{-13}{28}, \frac{-5}{14}$ (d) $\frac{205}{147}, \frac{-43}{147}$

Check 27

1 80, 80, 70   **2** 90   **3** 50, 50   **4** 80   **5** 30, 110, 30
6 124, 28, 34   **7** 55, 35, 55, 70   **8** 63, $31\frac{1}{2}$
9 $37\frac{1}{2}$, 75   **10** 120, 120

Check 28

4 (a) $\frac{1}{13}$ (b) $\frac{1}{3}$ (c) $\frac{4}{11}$ (d) $\frac{9}{25}$ (e) $\frac{3}{5}$
5 (a) $\frac{8}{25}$ (b) $\frac{3}{50}$ (c) $\frac{1}{10}$

Check 29

1 $\frac{5}{12}$ **2** $\frac{2}{5}, \frac{3}{5}, \frac{4}{5}$ **3** $\frac{3}{8}, \frac{1}{2}$
4 $\frac{1}{2}, \frac{2}{9}$ **5** $\frac{8}{27}, \frac{20}{27}$ **6** 0.14, 0.53

Check 30

1 18, 17   **2** -9 **4** (b) (c)
5 $\frac{1}{3}, \frac{2}{3}, \frac{1}{3}$ **6** 5, 13, 13

Check 31

1, 4, 5(b), 8

Check 32

1 (a) $18x^5$ (b) $3x^2 + 4x$ (c) $\dfrac{-12}{x^4}$ (d) $10x^4$ (e) $3 - \dfrac{2}{x^2}$

(f) $\dfrac{1}{\sqrt{x}}$ (g) $0.6x^{-0.8}$ (h) $\frac{1}{3}x^{-2/3} + \frac{3}{4}x^{-1/4}$ (i) $\dfrac{-x^{-3/2}}{6}$

(j) $2x - 2$ (k) $2x + 1$ (l) $6x^2(x^3 + 1)$ (m) $\dfrac{-1}{x^2}$

(n) $1 - \dfrac{1}{x^2}$ (o) $2x - 2 + \dfrac{2}{x^2}$

2 (a) $\dfrac{x^3}{3} + x + C$ (b) $x^3 - \dfrac{5x^2}{2} + x + C$ (c) $\tfrac{2}{3}x^{3/2} + C$

(d) $-\dfrac{1}{x} - \dfrac{1}{2x^2} + C$ (e) $\dfrac{(x-1)^3}{3} + C$ (f) $\dfrac{x^3}{3} - \dfrac{2}{x} + x^2 - \dfrac{2}{x^2} + C$

(g) $\dfrac{x^4}{4} - \dfrac{3x^2}{2} - \dfrac{1}{2x} + C$ (h) $\dfrac{x^7}{7} - x + C$

3 $5\tfrac{1}{3}$, $-4\tfrac{4}{5}$, $4\tfrac{2}{3}$, $1\tfrac{1}{3}$, $41\tfrac{1}{3}$, $6\tfrac{2}{3}$, $249\tfrac{3}{5}$, $1\tfrac{1}{2}$

Exercise 1

1 D   2 C   3 D   4 B   5 C   6 A   7 A   8 A   9 C   10 D
11 (a) {3, 7}   (b) {13, 15}
12 (a) {9, 36}   (b) {3, 16, 12, . . ., 33, 39, . . ., 60}
   (c) {4, 8, 16, 17, 25, 26, 35, 44, 49, 53}
13 62   14 16
15 (a) $(A \cap B \cap C') \cup (A \cap B' \cap C) \cup (A' \cap B \cap C)$
   (b) $(A \cup C) \cap B'$ (c) $((A \cup C)' \cap B) \cup ((A \cup B)' \cap C) \cup ((B \cup C)' \cap A)$
   (d) $((A \cup B)' \cap C) \cup (A \cap B \cap C')$
16 (a) ϕ (b) cyclic
17 (a) (i) S (ii) ϕ (b) (i) 6   (ii) 47
18 (a) There are no fast red cars   (b) Some red cars are fast
   (c) Some red cars are not fast
19 {5, 10, 15, 20, 25}, {3, 6, . . ., 27}, {5, 10, 20, 25}. . . .
20 (b) {3}   (c) {1, 4, 6, 9, 12, 15, 18, 21, 24, 27, 30, 33, 36, 39}.
21 (d) {2, 3, 13, 17, 19} (e) {1, 14, 15, 16, 18, 20}
   (f) {1, 6, 8, 9, 12, 14, 15, 16, 18}
22 (a) 19   (b) 32
23 (a) 32   (b) 44   (c) 3   (d) 20   (e) 4
24 (a) {multiples of 70} (b) 83
25 14   26 (a) 10   (b) 3   (c) 15   (d) 59
28 (a) (i) {3}   (ii) {1, 2, 3}   (iii) {1, 2, 3, 4, 5}
   (b) (i) {4, 16, 36}   (ii) {25, 36, 49}   (iv) {36}
29 (b) (ii) 24   (iii) 3

Exercise 2

1 C   2 B   3 A   4 D   5 A   6 A   7 B   8 D   9 B   10 A   11 D
12 C   13 C   14 £30.17   15 £288 000, £300, 82p
16 (a) 9, 6 (b) $3n + 4$ 17 0.02   18 £87.72   19 (a) 101010   (b) 110
(c) 111000011   (d) 100011   21 (a) 6   (b) 4   (c) 8
22 7.44×10^5, -2.08×10^5 23 0, 36   24 2
25 (a) 2.4393   (b) 0.141   (c) 22.41   26 9%
27 (a) £75 000   (b) (i) 3.2p   (ii) £17.28   (c) (i) £230,   (ii) 2.7%
28 (a) (i) £159   (ii) 223   (iii) 205   (b) £624   (c) £245.30
29 (a) £104.12$\tfrac{1}{2}$ (b) £178.50   (c) 6.25%   (d) £233.50
30 187, 40   31 9.46×10^{12} km, 2.37×10^{14} km
32 (a) 2361.6   (b) 85 000   (c) 125   (d) 1676
33 (a) 2660   (b) 9.5, £434, 2.2, £84

Exercise 3

1 D **2** C **3** D **4** B **5** C **6** B **7** C **8** A **9** C **10** B **11** D **12** A

13 ± 6 **14** $by^2/(x-1)$ **15** $(x-3)(2x+1)(x-1)$

16 $(2a+4b+5c)(2a+4b-5c)$ **17** $(v^2-u^2)/2s$ **18** (a) $\sqrt{29}$ (b) $\frac{1}{3}$

19 -1 **20** (a) 19 (b) $-5\frac{1}{2}$ (c) $-\frac{4}{5}$

21 (a) 200.016 (b) 1.96 (c) $\sqrt{\dfrac{k}{T-100n}}$

22 (a) $(x+1)(4x+1)$ (b) $4(3a+b)(2b-a)$ (c) $x(x+3)(x-2)$

23 $\phi, \frac{1}{3}, \sqrt{2}$ (d) $\sqrt{2}, \frac{1}{3}$ **24** (a) $2p^2$ (b) $8x$ (c) $18x^4$

25 (a) $x-1$ (b) $x, \dfrac{1}{(x-1)}$ (c) 1.62

26 (a) $\sqrt{\dfrac{qy^2-3t}{4t}}$ (b) $\dfrac{-(ky+m)}{h}$ (c) $\dfrac{yt}{4+k}$

27 (a) (i) $17\frac{1}{2}$ (ii) 2 (b) 1.77

28 (a) 29 (b) $2nx-P$ (c) $\dfrac{P+y}{2x}$ (d) no effect

29 (a) $\frac{7}{8}$ (b) $-2, 1\frac{1}{2}$ (c) $0, -3$ (d) $2\frac{1}{7}$ (e) 1 (f) ± 5 (g) $-\frac{1}{4}$ (h) -1
(i) ± 3 (j) 2

30 (a) $q-4y$ (b) $(4-2T^2)/T^2$ (c) $t/(y-4)$ (d) $\sqrt{r^2-y^2}$ (e) $\sqrt{4+y^2}$
(f) $\sqrt{\dfrac{A^2t^2-16y^2}{16}}$ (g) $-(by+c)/a$ (h) $\sqrt{\dfrac{2E}{m}}$ (i) $\dfrac{f}{Ff-1}$

31 (a) $2t(q-3)(q+3)$ (b) $4t(q+2t)$ (c) $(2t-3)(x+y)$
(d) $(8x+11)(4x-9)$ (e) $(x-1)(x^2+2x+2)$

32 (a) 4 (b) 13 (c) 9 (d) -38 **33** (a) $\dfrac{50}{x^2}$ (b) $4x^2+\dfrac{300}{x}$

34 (a) $2, -1$ (b) $vf/(v-f)$ (c) $(x+4)^2+9$

35 $(wy+(52-w)r)/52$ **36** 6.52

37 (a) 1600 (b) 8   **38** 366

39 (a) (ii) $(A+b^2)/2b$ (b) $4l$ **40** $1.74 \times 1.74 \times 4.74$

41 240, 2 h 5 min

Exercise 4

1 B **2** A **3** D **4** A **5** B **6** B **7** A **8** D **9** B **10** D **11** $\frac{8}{27}$ cm

12 64 cm^2 **13** (a) 8.8×10^5 (b) 2.72×10^{-2} **14** (a) 90 (b) 90

15 £982 600, 520 km

16 (a) £6000, £3600, £2400 (b) 370, £4500, 75%

17 (a) 2240 m^3 (b) 1640 cm^3 (c) $\frac{70}{33}$

18 (a) 5960 mm (b) £178.80 (c) £208.60 (d) £47.25 (e) £434.65
(f) £43.46$\frac{1}{2}$ (g) £23.91

19 (c) 52   **20** 48 m^3, 5 min, 37.5 cm

21 (c) (ii) 8 m/s^2 **22** (a) 60 (b) 11.3 (c) $3\frac{3}{4}$ m/s^2

23 $3\frac{1}{3}$ min, 67 min   **24** 83.4 cm^3, 8.2%, 1.5 cm

Exercise 5

1 D **2** B **3** B **4** A **5** C **6** D **7** B **8** D **9** D **10** D **11** C

14 (a) -0.74 (b) 0.66   **15** (a) 2 (b) $(-1\frac{1}{2}, 3)$ (c) $(3, -6)$

16 (a) (ii) $\frac{11}{5}$ (b) $y+2x+3=0$ **17** $x^3+1.6x^2+x+1=0$

18 (a) $x = 2.24$ (b) 3.3 or 1.5   19 (c) (i) 0.81   (ii) -4.1 (iii) 11.3
                                            (e) 1, $-2\frac{1}{3}$
20 2.4, 0.6   21 (a) 6 (b) $6\sqrt{3}, 2\sqrt{3}$ (c) 11.6, 191.6
22 (a) (10, 0) (0, 5) (b) -0.5 (c) $y = 2x$ (d) (2, 4) (e) (4, 8) (f) (5, $2\frac{1}{2}$)
   (g) $5\sqrt{5}/2$ 23 (a) $-\frac{1}{60}$ (b) 0, $(-4, -5)$ (c) 1.93
24 (a) (i) $x \mapsto 2$ (ii) $x \mapsto 4$ (b) gh (c) $x \mapsto 3(x - 1)$
25 (a) 2 (b) 0, 1, 2 (c) 0
26 (a) $(3 - x)/2$ (b) $2/3x$ (c) $3\sqrt{x}$ (d) $3\sqrt{x - 2}$ (e) $(1 - x)/2x$
   (f) $(b - dx)/(cx - a)$ 28 4 solutions, 2.65
29 (a) -22 (b) -2 (c) -22 (d) $-4, 9$ (e) 9.92, -1.92
30 (a) 18.8 (b) ± 5.29 (c) $-\frac{7}{4}$ (d) $y \geqslant 6$
31 (a) 2.5 (b) 3.6 (c) $2.5 < x < 4.1$

Exercise 6

1 D   2 C   3 A   4 D   5 A   6 D   7 C   8 B   9 D   10 C   11 8, 9, 10
12 (a) 24   (b) (i) $2\frac{1}{4}$ (ii) 0   13 (c) $5x + 3y + 55$ (d) 125
14 £4.80   15 $x + y \leqslant 1000, y \geqslant 2x, x \geqslant 100, y \leqslant 800$, 333, 667
16 (a) $h \geqslant 15$ (b) $p > 25$ (c) $45 \leqslant h + p < 60$; 25. 30

Exercise 7

1 D   2 B   3 C   4 B   5 C   6 C   7 C   8 B   9 A   10 D   11 (a) 34   (b) 26
12 7, 4   13 $5\overrightarrow{PT}$ 14 0, 6
15 (a) (i) $2\mathbf{v}$ (b) $3\mathbf{v} + 2\mathbf{u}$ (c) $-2\mathbf{v} - 2\mathbf{u}$ (d) $\overrightarrow{QN}$
16 (d) $\frac{2}{3}$ 17 (d) 2.6 km   (e) 12.16
18 $h = k = \frac{1}{3}$, 1 : 3   19 (b) $\frac{1}{3}$ (c) $\frac{21}{16}$ (d) $\frac{4}{9}$
20 (c) $(\mathbf{a} + \mathbf{b} + \mathbf{c} + \mathbf{d})/4$
21 (a) $10\mathbf{a} - 2\mathbf{b}$ (b) $2\mathbf{b} + k(10\mathbf{a} - 2\mathbf{b})$ (c) $10m\mathbf{a} + (2\mathbf{b} - 10\mathbf{a})$
22 (a) (i) $-\mathbf{x}$ (ii) $2\mathbf{x}$ (iii) $\mathbf{y} - \mathbf{x}$ (iv) $\mathbf{x} + \mathbf{y}$ (iv) $2\mathbf{x} - \mathbf{y}$ (b) (i) $2\mathbf{x}$ (ii) $4\mathbf{x} - 2\mathbf{y}$
   (c) (i) 2   (ii) 6   24 (a) (i) $2\mathbf{n} + 2\mathbf{e}$ (ii) $2\sqrt{2}$ (b) $k = \frac{2}{5}$
25 (a) (i) $\begin{pmatrix} -6 \\ -2\frac{1}{2} \end{pmatrix}$ (ii) 13   (b) $3\mathbf{a} + 3\mathbf{b}, -\mathbf{a} + 3\mathbf{b}, m = 4, n = 3$

Exercise 8

1 B   2 D   3 C   4 A   5 D   6 D   7 B   8 1 : 8   12 (3, 1), (2, 3)

13 (6, 4), H 14 (a) $\begin{pmatrix} 8 \\ 4 \end{pmatrix}, 4\sqrt{5}$ (b) (10, 8) (c) $-1\frac{1}{2}$

15 90° anticlockwise about k 16 (b) $-90°$ (c) $\begin{pmatrix} 2 \\ -2 \end{pmatrix}$
                                           (d) $-90°$ about $(0, -2)$

17 (a) $D (-2, -2)$
18 (a) similar (b) equal (c) $\frac{3}{4}$ (d) 1 : 16 (e) 1 : 4 (f) 15 cm²
19 $\frac{3}{4}$

Exercise 9

1 A 2 C 3 D 4 D 5 A 6 D 7 D 8 A 9 D 10 C 11 $\sqrt{\frac{3}{2}}, \sqrt{\frac{3}{2}}, 3$

12 (1) $\begin{pmatrix} 3 & 6 \\ -1 & -2 \end{pmatrix}$　　14 $\begin{pmatrix} 15 \\ 12 \end{pmatrix}$　　15 $\begin{pmatrix} 0 & 2 \\ 0 & 2 \end{pmatrix}, \begin{pmatrix} -1 & -2 \\ 0 & 1 \end{pmatrix},$

reflection in $x = 0$　　　　　　17 18, -11

18 8, 24　　　　　　　　　　　19 rotation 53° clockwise

20 $(2, 1), (3\frac{2}{3}, -\frac{1}{3}), (-\frac{1}{3}, -\frac{1}{2})$

22 (a) straight line　(b) parallel line × 2　(c) enlargement × 4

23 1, -1　　　　　　　　　　24 $d^3 + e^3 d + e^2 d^2 - e^2 d$

25 $\begin{pmatrix} 0 & -\frac{1}{2} \\ 1 & 0 \end{pmatrix}\begin{pmatrix} -1 & 0 \\ 0 & -1 \end{pmatrix}$

27 (b) (ii) $\begin{pmatrix} 5 & 0 & 0 \\ 0 & 5 & 0 \\ 0 & 0 & 4 \end{pmatrix}$　　　28 (a) (i) $\begin{pmatrix} 7 & 0 \\ 0 & -7 \end{pmatrix}$ (ii) $\begin{pmatrix} 6 & -3 \\ -10 & -16 \end{pmatrix}$

29 $t = 2$　　　　30 $x_n^2 = y_n y_{n-1} + 1,$ $\begin{pmatrix} 2 & -1 \\ -1 & 1 \end{pmatrix}, \begin{pmatrix} 2 & 3 \\ 3 & 5 \end{pmatrix}$

32 (a) (i) (3, 9)　(ii) (14, 8)　(iii) (2, -8)　(b) (i) -3　(ii) $-1, 2$

Exercise 10

1 (a) 20°　(b) 10.3 cm　(c) 20.3 cm　2 $52\frac{1}{2}°, 127\frac{1}{2}°, 82\frac{1}{2}°, 97\frac{1}{2}°$

3 13 cm　5 140°　　　　　　　6 (a) 60°　(b) 240°　(c) 30°

7 (a) 64°　(b) 107°　　　　　　8 100°

9 POT, AQO, QTC　　　　　　10 (a) $180 - x$　(b) $360/x$

Exercise 11

1 B 2 C 3 B 4 C 5 D 6 B 7 B 8 A 9 C 10 C 11 $\frac{1}{2}, \frac{1}{5}, \frac{1}{10}$

12 C   13 $533\frac{1}{3}$ 14 $\frac{1}{10}$ 15 $\frac{19}{36}$ 16 6   17 $\frac{1}{3}$ 18 $\frac{1}{6}, \frac{5}{6}, \frac{1}{30}$ 19 $\frac{7}{494}, \frac{3}{200}$

20 $\frac{7}{9}, \frac{1}{25}, \frac{17}{75}$ 21 (a) $\frac{1}{2}$ (b) (i) $\frac{1}{4}$ (ii) $\frac{1}{4}$ 22 (a) $\frac{15}{16}$ (b) $12\frac{1}{2}$

23 (a) 6　(b) 12　(c) 12　(d) £5600　24 (a) 0.343　(b) 0.1029　(c) 0.3087
　　　　　　　　　　　　　　　　　　　　　(d) 0.6517

25 0.24, 600   26 (a) (i) $\frac{1}{36}$ (ii) $\frac{1}{6}$ (b) (i) $\frac{4}{25}$ (ii) $\frac{41}{100}$ (c) 0.202

27 (b) 114　(c) 0.316, 0.673, 0.010

28 (a) $\frac{1}{3}$ (b) (i) $\frac{4}{9}$ (ii) $\frac{1}{6}$　　29 (b) 48, 49　(c) 49.1

Exercise 12

1 B 2 A 3 A 4 D 5 C 6 D 7 D 8 C 9 D 10 C

11 7.38 cm, 49.5°, 14.9 cm　　　12 20.5 cm, 6.43 cm, 2, 63.4°

13 (a) 7.2, 10.8　(c) 56.3°　(d) 55.3°　14 25.1 m, 65°

15 13.6 cm　　　　　　16 $-\frac{2}{3}$　17 123.7°, $\frac{7}{25}$

18 144.7°, 60.2 n miles, 78.3°　　19 (a) (i) 254.8 m　(ii) 27°　(b) 331.5 m

20 (a) 4.8 cm　(b) 9.6 km　(c) 45.4°

21 (a) 33.4°, 146.6°　(b) (i) 6.65 miles　(ii) 3.14 miles

22 (a) 173.8 m　(b) 337°　　　24 (a) 14°　(b) 4.84 m

25 (a) 10 cm　(b) 200 cm³　　　26 (a) 84.3°　(b) 151.4 cm　(c) 81.2°

27 (a) 3119 m　(b) 3589 m　(c) 3569 m　28 6.4 m, 9.38 m²

29 7.14 cm   **30** (a) 8.29 cm   (b) 14.9 cm   (c) 17.4 cm²   (d) 2.34 cm
31 4.68 cm, 3.96 cm   **32** 53.5°, 126.5°, 3.21 cm
33 (a) 14.4   (b) 46.2°   (c) 0.87 km   **34** 190.4 m, 248 m, 13.5°
35 8.65 cm, 5.47 cm, 30.6 cm²   **36** $23.7 \leqslant y \leqslant 110.1$
37 5.45 cm   **38** (a) 8.38   (b) 25.13   (c) 17.73   (d) 7.71
39 (a) 2.24 m   (b) 2.84 m   (c) 120° 4.1 m
40 (56° S, 143° E)   **41** 75.5°
42 (a) 2751 km   (b) 8642 km, 3070 n miles
43 (a) 10 cm   (b) 8.66 cm   (c) $\frac{5}{6}$ (d) 23.6° N
44 (a) 7440 n miles   (b) 9536 n miles, (28° N, 21.6° W)
45 (a) (i) 37   (ii) 53   (iii) 106   (iv) 74   (b) (i) $f = 180 - 2x$,   (ii) $g = 2x$
(iii) all values

Exercise 13

1 (a) perpendicular line   (b) parallel line
3 $CQ = 8.1$ cm   **7** 3.5 cm   **12** 126 cm³   **14** 275 m²

Exercise 14

1 B   **2** C   **3** A   **4** E   **5** C   **6** B, D   **7** D   **8** A   **9** C   **10** C   **11** (a) 3   (b) 4
12 (a) r (b) p (c) yes   **13** (a) 12   (b) 1   **14** (a) (c) (e) closed
15 (c) (d) (f) (g)   **16** (a) (b) (e) (f)   **17** (a) d (b) b (c) c
18 (c) −2   **19** (a) (i) 5   (ii) 3   (b) 7, 8

Exercise 15

1 A   **2** B   **3** C   **4** D   **5** B   **6** D   **7** A   **8** B   **9** B   **10** C
11 (a) $(-\frac{1}{6}, -4\frac{1}{12})$ min.   (b) −6.35   (c) $x = -\frac{1}{6}$
12 (a) 27 m   (b) 1.5 m/s   **13** (a) (2, 4)   (b) $10\frac{2}{3}$ (c) 16π
14 (a) $40\frac{1}{3}$ (b) $\frac{37}{24}$ (c) $-\frac{4}{3}$
16 (a) 101.25 m   (b) 5 s   (c) 45 m/s
18 (a) $y + 8x + 3 = 0$, (1, 17) max.   (b) (6, −108) min.
19 (a) (i) −1, 4   (ii) 9   (b) (ii) 1   (iii) 80/3
20 (b) (i) 3   (ii) 14 cm from X (c) $0 < t < \sqrt{\frac{2}{3}}$ (e) 28
21 (a) 27 cm   (b) 14 cm/s   (c) −14 cm/s²   (d) $1\frac{1}{2}$ s   (e) 35 cm
22 (a) 32 m/s   (b) −12 m/s²   (c) $15\frac{2}{3}$ m   **23** (a) $0.4x^8 + 2/x$ (b) 2, −3
24 (a) (ii) − 15   (iii) −3   (iv) $\frac{5}{3}$ min., −3 max.   **25** (a) 5   (b) −15   (c) (4, 12)
                                                                                                (d) 30.5
26 (a) (2, 2)   (c) $3\frac{1}{6}$ (d) 1.78   **27** (a) 2   (c) $(\frac{1}{2}, \frac{3}{4})$ (d) 3.35
28 (b) 4   (c) (3, −9)   (e) $14\frac{2}{3}$

Index

Acceleration 77, 249
Algebraic expressions 48
Algebraic fractions 47
Alternate angle 159
Approximation 28
Area 66, 251
Arithmetic mean 181
Associative 40
Assumed mean 183

Bar chart 179
Base 18
Bearings 197
Binary operation 234

Change of subject 54
Chord 165
Circle 67, 90
Circumscribing circle 222
Closure 234
Codomain 85
Commutative 40
Complement 3
Composite function 86
Compound interest 25
Cone 68
Congruence 128
Constructions 220
Corresponding angles 159
Cosine 190
Cosine rule 203
Cuboid 68
Cumulative frequency 182
Curve sketching 96
Cylinder 68

Decimal places 18
Definite integral 251
De Morgan's laws 6
Density 68
Depression 197
Determinant 141
Difference of two squares 45
Differentiation 246
Dilatation 126
Directed numbers 39
Disjoint sets 2
Distance 90
Distance–time graph 77, 79
Distributive laws 6, 40
Domain 85

Element 1
Elevation 197
Enlargement 126, 149
Equations 49
    forming 52
Errors 103
Euler's theorem 146
Exchange rates 24
Exponential 96

Factorization 44
Fractions 21
Frequency 181
Functions 85

Glide reflection 126
Gradient 89, 94, 246
Great circle 207
Group 236

HCF 46
Histogram 184
Hyperbola 96

Identity matrix 141
Image 85
Incidence matrix 145
Income tax 33
Indices 42
Inequalities 102
Inequations 102
Infinite set 2
Inscribed circle 223
Integration 250
Intersection 3
Inverse function 86
Inverse matrix 141
Irrational numbers 17
Isometry 126
Isomorphism 241

Kite 164

Latitude 207
LCM 46
Like terms 43
Linear equation 49
Linear programming 105

Locus 109, 118, 229
Logarithms 30
Logic 5
Longitude 207

Matrices 139
Maxima and minima 254
Median 181
Member 2
Metric system 65
Mid-point 90
Mirror line 124
Mode 181
Modulo arithmetic 241
Modulus 114

Nautical mile 207
Network 144
Normal 248
Null set 3
Number system 17

Ogive 182
Operation table 235
Order 161

Parabola 96
Parallelogram 67, 163
Percentages 22
Perfect square 46
Pictogram 180
Pie chart 180
Plan and elevation 227
Polygon 160
Prism 68
Probability 173
Proportion 22, 57, 96
Pyramid 68
Pythagoras 192

Quadratic equation 50

Radian 74
Range 85
Rates 32
Rational numbers 17
Rectangle 67, 163
Reflection 124, 147

Region   104
Remainder theorem   56
Rhombus   67, 164
Rotation   125, 148

Sample space   173
Scalar   113, 139
Scale   24
Secant   166
Sector   75, 165
Segment   75, 165
Semi-interquartile range   183
Sets   2
Shear   127, 149
Significant figures   18
Similarity   128, 133
Simple interest   25
Simultaneous equations   51, 142
    Non-linear   58
Sine   190
Sine rule   201

Singular   141
Speed   77
Sphere   68
Standard form   26
Straight line   89
Stretch   128, 150
Subgroup   241
Subject of formula   54
Subset   4
Supplementary   159
Surds   29
Surface area   68
Symmetry   161

Tangent   190, 248
Transformations   147, 150
Translation   124
Trapezium   67, 164
Trapezium rule   73, 78
Traversability   146

Tree diagram   176
Triangle   67, 162

Union   3
Units   65
Universal set   3

Variation   57, 96
Vector equation   115
Vectors   113
Velocity   77, 249
Velocity-time graph   77
Velocity triangle   118, 209, 223
Venn diagram   4
Vertically opposite   159
Volume   68
Volume of revolution   253

Wages   34